Version 2

AF173957

College
Algebra
+ Integrated Review

Guided Notebook

 HAWKES
LEARNING

Editors:
Robin Hendrix,
S. Rebecca Johnson,
Barbara Miller,
Amber Widmer

Copy Editors:
Mary Janelle Cady,
Rebecca Lebeaux

Creative Director:
Tee Jay Zajac

Designers:
Natalie Ezabele,
Trudy Gove,
D. Kanthi,
E. Jeevan Kumar,
U. Nagesh,
B. Syam Prasad,
Patrick Thompson

Cover Design:
Trudy Gove

Answer Key Editors:
Y.Appaji,
G.S.V.Bhargav,
P.Harika,
K.Karteek,
U.Ravindra Kumar,
K.Sandeep Kumar,
G.Sateesh Kumar,
S.V.Anil Kumar Manda,
K.Ramesh Narasimha,
A.Navyasri,
Ch.Prasad,
M.V.Kiran Prasad,
K.N.V.Madhavi Prasanna,
K.Sitarama Raju,
N.Narasimha Rao,
G.Ramjagadish,
V.Ravi Sankar,
K.Ch.Sindhusha

VP Research & Development: Marcel Prevuznak

Director of Content: Kara Roché

A division of Quant Systems, Inc.

546 Long Point Road, Mount Pleasant, SC 29464

Printed in the United States of America

1 0 9 8 7 6 5 4 3 2 1

ISBN: 978-1-64277-094-0

Table of Contents

Strategies for Academic Success

CHAPTER 1.R

Review Concepts
for *Number Systems and Fundamental Concepts of Algebra*

CHAPTER 2.R

Review Concepts
for *Equations and Inequalities of One Variable*

Math@Work

Answer Key

Math Knowledge Required for Math@Work Career Explorations

The following table summarizes the math knowledge required for each Math@Work career exploration. Use this table to determine when you are ready to explore each career.

Math@Work Career	Whole Numbers	Fractions	Integers	Decimal Numbers	Averages	Percents	Simple Interest	Ratios	Proportions	Geometry	Statistics	Graphing	Linear Equations	Systems of Equations	Mixture Problems	Scientific Notation	Greatest Common Factor	Rational Expressions	Radicals
Basic Inventory Management	✓																		
Hospitality Management	✓	✓			✓														
Bookkeeper				✓															
Pediatric Nurse				✓				✓	✓										
Architecture				✓						✓									
Statistician: Quality Control				✓							✓	✓							
Dental Assistant				✓	✓														
Financial Advisor				✓		✓	✓						✓						
Market Research Analyst				✓									✓						
Chemistry				✓											✓	✓			
Astronomy				✓												✓			
Math Education			✓														✓		
Physics		✓																✓	
Forensic Scientist				✓	✓														✓
Other Careers in Mathematics																			

✇ Support

If you have questions or comments we can be contacted as follows:

24/7 Chat: chat.hawkeslearning.com

Phone: (843) 571-2825

E-mail: support@hawkeslearning.com

Web: hawkeslearning.com

Our support hours are 8:00 a.m. to 10:00 p.m. (ET), Monday through Friday.

How to Use the Guided Notebook

There are a variety of elements in this Guided Notebook that will help you on your way to mastering each topic. Here is a rundown of how to use the elements as you work through this notebook.

Fill-in-the-Blanks

1. When there is an incomplete sentence, you will need to write in the _____*missing*_____ word(s).

   The _____*missing*_____ words can be found by reading through the Learn screens.

Boxed Content

Definitions and **procedures** are highlighted within a box, like the ones shown here. The missing content will vary from box to box. Sometimes an entire definition is missing and sometimes only part of a sentence is missing. Here are two examples of the box variations.

Definition

First term to define: _____*Write the definition here.*_____

Second term to define: _____*If there is another term, define it the same way as above.*_____

DEFINITION

Terms Related to Probability

_____*Outcome*_____	An individual result of an experiment.
_____*Sample Space*_____	The set of all possible outcomes of an experiment.
_____*Event*_____	Some (or all) of the outcomes from the sample space.

DEFINITION

Properties and **Procedure** boxes are completed in a similar way:

> ## Commutative Property of Multiplication
> The order of the numbers in multiplication can be _reversed without changing the product._
>
> For example, _$3 \cdot 4 = 12$ and $4 \cdot 3 = 12$._
>
> **PROPERTIES**

> ## Subtracting Whole Numbers
> 1. Write the numbers _vertically_ so that the _place values are lined up in columns._
>
> 2. Subtract only the _digits with the same place value._
>
> 3. Check by _adding the difference to the subtrahend._ The sum must be _the minuend._
>
> **PROCEDURE**

▶ Watch and Work

For each Watch and Work, you will need to watch the corresponding video in Learn mode and follow along while completing the example in the space provided.

Example 5 Multiplying Whole Numbers

Multiply: $12 \cdot 35$

Solution

The standard form of multiplication is used here to find the product $12 \cdot 35$.

$$
\begin{array}{r}
\overset{1}{1}2 \\
\times\ 35 \\
\hline
60 \\
360 \\
\hline
420
\end{array}
$$

$12 \cdot 5 = 60$
$12 \cdot 30 = 360$
Product

✏ Now You Try It!

After working along with the example video, work through a similar exercise on your own in the space provided.

Example A Multiplying Whole Numbers

Multiply:
$$
\begin{array}{r}
25 \\
\times\ 42 \\
\hline
1050
\end{array}
$$

1.1 Exercises

Each section has exercises to offer additional practice problems to help reinforce topics that have been covered. The exercises include Concept Check, Practice, Application, and Writing & Thinking questions. The odd answers can be found in the Answer Key at the back of the book.

Concept Check

True/False. Determine whether each statement is true or false. If a statement is false, explain how it can be changed so the statement will be true. (**Note:** There may be more than one acceptable change.)

1. When the given statement is true, you write "True" for the answer.

   True

Practice

For each set of data, find **a.** the mean, **b.** the median, **c.** the mode (if any), and **d.** the range.

2. *Presidents:* The ages of the first five US presidents of the 20th century on the date of their inaugurations were as follows. (The presidents were Roosevelt, Taft, Wilson, Harding, and Coolidge.)

   42, 51, 56, 55, 51

   a. 51 b. 51 c. 51 d. 14

Applications

Solve.

3. *Grades:* Suppose that you have taken four exams and have one more chemistry exam to take. Each exam has a maximum of 100 points and you must average between 75 and 82 points to receive a passing grade of C. If you have scores of 85, 60, 73, and 76 on the first four exams, what is the minimum score you can make on the fifth exam and receive a grade of C?

   81

Writing & Thinking

4. State how to determine the median of a set of data.

   The first step to finding the median is always to arrange the data in order. Once the data is in order, the median is the number in the middle. If there is an even number of items, average the two middle numbers to find the median.

Strategies for Academic Success

Strategies for Academic Success 🎓

How to Read a Math Textbook

Reading a textbook is very different than reading a book for fun. You have to concentrate more on what you are reading because you will likely be tested on the content. Reading a math textbook requires a different approach than reading literature or history textbooks because the math textbook contains a lot of symbols and formulas in addition to words. Here are some tips to help you successfully read a math textbook.

Don't Skim 📖

When reading math textbooks, look at everything: titles, learning objectives, definitions, formulas, text in the margins, and any text that is highlighted, outlined, or in bold. Also pay close attention to any tables, figures, charts, and graphs.

Minimize Distractions

Reading a math textbook requires much more concentration than a novel by your favorite author, so pick a study environment with few distractions and a time when you are most attentive.

🚩 Start at the Beginning

Don't start in the middle of an assigned section. Math tends to build on previously learned concepts and you may miss an important concept or formula that is crucial to understanding the rest of the material in the section.

Highlight and Annotate

Put your book to good use and don't be afraid to add comments and highlighting. If you don't understand something in the text, reread it a couple of times. If it is still not clear, note the text with a question mark or some other notation so you can ask your instructor about it.

Go through Each Step of Each Example ≣

Make sure you understand each step of an example. If you don't understand something, mark it so you can ask about it in class. Sometimes math textbooks leave out intermediate steps to save space. Try working through the examples on your own, filling in any missing steps.

Take Notes *< This is important!*

Write down important definitions, symbols or notation, properties, formulas, theorems, and procedures. Review these daily as you do your homework and before taking quizzes and tests. Practice rewriting definitions in your own words so you understand them better.

Notes 9-25-17:

- The opposite of a negative integer is a positive integer.

- To add two integers with the same signs add their absolute values and use their common sign

🖥 Use Available Resources

Many textbooks have companion websites to help you understand the content. These resources may contain videos that help explain more complex steps or concepts. Try searching the internet for additional explanations of topics you don't understand.

Read the Material Before Class

Try to read the material from your book before the instructor lectures on it. After the lecture, reread the section again to help you retain the information as you look over your class notes.

Understand the Mathematical Definitions + × =

Many terms used in everyday English have a different meaning when used in mathematics. Some examples include equivalent, similar, average, median, and product. Two equations can be equivalent to one another without being equal. An average can be computed mathematically in several ways. It is important to note these differences in meaning in your notebook along with important definitions and formulas.

Try Reading the Material Aloud

Reading aloud makes you focus on every word in the sentence. Leaving out a word in a sentence or math problem could give it a totally different meaning, so be sure to read the text carefully and reread, if necessary.

Questions

1. Explain how taking notes can help you understand new concepts and skills while reading a math textbook.

2. Think of two more tips for reading a math textbook.

Strategies for Academic Success 🎓

Tips for Success in a Math Course

Read Your Textbook/Workbook

One of the most important skills when taking a math class is knowing how to read a math textbook. Reading a section before class and then reading it again afterwards is an important strategy for success in a math course. If you don't have time to read the entire assigned section, you can get an overview by reading the introduction or summary and looking at section objectives, headings, and vocabulary terms.

Take Notes ✏️

Take notes in class using a method that works for you. There are many different note-taking strategies, such as the Cornell Method and Concept Mapping. You can try researching these and other methods to see if they might work better than your current note-taking system.

Review

While the information is fresh in your mind, read through your notes as soon as possible after class to make sure they are readable, write down any questions you have, and fill in any gaps. Mark any information that is incomplete so that you can get it from the textbook or your instructor later.

📁 Stay Organized

As you review your notes each day, be sure to label them using categories such as definition, theorem, formula, example, and procedure. Try highlighting each category with a different colored highlighter.

Use Study Aids

Use note cards to help you remember definitions, theorems, formulas, or procedures. Use the front of the card for the vocabulary term, theorem name, formula name, or procedure description. Write the definition, the theorem, the formula, or the procedure on the back of the card, along with a description in your own words.

Practice, Practice, Practice!

Math is like playing a sport. You can't improve your basketball skills if you don't practice—the same is true of math. Math can't be learned by only watching your instructor work through problems; you have to be actively involved in doing the math yourself. Work through the examples in the book, do some practice exercises at the end of the section or chapter, and keep up with homework assignments on a daily basis.

▦ Do Your Homework

When doing homework, always allow plenty of time to finish it before it is due. Check your answers when possible to make sure they are correct. With word or application problems, always review your answer to see if it appears reasonable. Use the estimation techniques that you have learned to determine if your answer makes sense.

Understand, Don't Memorize

Don't try to memorize formulas or theorems without understanding them. Try describing or explaining them in your own words or look for patterns in formulas so you don't have to memorize them. For example, you don't need to memorize every perimeter formula if you understand that perimeter is equal to the sum of the lengths of the sides of the figure.

Study

Plan to study two to three hours outside of class for every hour spent in class. If math is your most difficult subject, then study while you are alert and fresh. Pick a study time when you will have the least interruptions or distractions so that you can concentrate.

⏱ Manage Your Time

Don't spend more than 10 to 15 minutes working on a single problem. If you can't figure out the answer, put it aside and work on another one. You may learn something from the next problem that will help you with the one you couldn't do. Mark the problems that you skip so that you can ask your instructor about it during the next class. It may also help to work a similar, but perhaps easier, problem.

Questions

1. Based on your schedule, what are the best times and places for you to study for this class?

2. Describe your method of taking notes. List two ways to improve your method.

Strategies for Academic Success 🎓

Tips for Improving Math Test Scores

Preparing for a Math Test

- Avoid cramming right before the test and don't wait until the night before to study. Review your notes and note cards every day in preparation for quizzes and tests.

- If the textbook has a chapter review or practice test after each chapter, work through the problems as practice for the test.

- If the textbook has accompanying software with review problems or practice tests, use it for review.

- Review and rework homework problems, especially the ones that you found difficult.

- If you are having trouble understanding certain concepts or solving any types of problems, schedule a meeting with your instructor or arrange for a tutoring session (if your college offers a tutoring service) well in advance of the next test.

Test-Taking Strategies

- Scan the test as soon as you get it to determine the number of questions, their levels of difficulty, and their point values so you can adequately gauge how much time you will have to spend on each question.

- Start with the questions that seem easiest or that you know how to work immediately. If there are problems with large point values, work them next since they count for a larger portion of your grade.

- Show all steps in your math work. This will make it quicker to check your answers later once you are finished since you will not have to work through all the steps again.

- If you are having difficulty remembering how to work a problem, skip it and come back to it later so that you don't spend all of your time on one problem.

After the Test

- The material learned in most math courses is cumulative, which means any concepts you miss on each test may be needed to understand concepts in future chapters. That's why it is extremely important to review your returned tests and correct any misunderstandings that may hinder your performance on future tests.

- Be sure to correct any work you did wrong on the test so that you know the correct way to do the problem in the future. If you are not sure what you did wrong, get help from a peer who scored well on the test or schedule time with your instructor to go over the test.

- Analyze the test questions to determine if the majority came from your class notes, homework problems, or the textbook. This will give you a better idea of how to spend your time studying for the next test.

- Analyze the errors you made on the test. Were they careless mistakes? Did you run out of time? Did you not understand the material well enough? Were you unsure of which method to use?

- Based on your analysis, determine what you should do differently before the next test and where you should focus your time.

Questions

1. Determine the resources that are available to you to help you prepare for tests, such as instructor office hours, tutoring center hours, and study groups.

2. Discuss two additional test taking strategies.

Strategies for Academic Success 🎓

Practice, Patience, and Persistence!

Have you ever heard the phrase "practice makes perfect"? This saying applies to many things in life. You won't become a concert pianist without many hours of practice. You won't become an NBA basketball star by sitting around and watching basketball on TV. The saying even applies to riding a bike. You can watch all of the videos and read all of the books on riding a bike, but you won't learn how to ride a bike without actually getting on the bike and trying to do it yourself. The same idea applies to math. Math is not a spectator sport.

Math is not learned by sleeping with your math book under your pillow at night and hoping for osmosis (a scientific term implying that math knowledge would move from a place of higher concentration—the math book—to a place of lower concentration—your brain). You also don't learn math by watching your professor do hundreds of math problems while you sit and watch. Math is learned by doing. Not just by doing one or two problems, but by doing many problems. Math is just like a sport in this sense. You become good at it by doing it, not by watching others do it. You can also think of learning math like learning to dance. A famous ballerina doesn't take a dance class or two and then end up dancing the lead in The Nutcracker. It takes years of practice, patience, and persistence to get that part.

Now, we aren't suggesting that you dedicate your life to doing math, but at this point in your education, you've already spent quite a few years studying the subject. You will continue to do math throughout college—and your life. To be able to financially support yourself and your family, you will have to find a job, earn a salary, and invest your money—all of which require some ability to do math. You may not think so right now, but math is one of the more useful subjects you will study.

It's important not only to practice math when taking a math course, but also to be patient and not expect immediate success. Just like a ballerina or NBA basketball star, who didn't become exceptional athletes overnight, it will take some time and patience to develop your math skills. Sure, you will make some mistakes along the way, but learn from those mistakes and move on.

Practice, patience, and persistence are especially important when working through applications or word problems. Most students don't like word problems and, therefore, avoid them. You won't become good at working word problems unless you practice them over and over again. You'll need to be patient when working through word problems in math since they will require more time to work than typical math skills exercises. The process of solving word problems is not a quick one and will take patience and persistence on your part to be successful.

Just as you work your body through physical exercise, you have to work your brain through mental exercise. Math is an excellent subject to provide the mental exercise needed to stimulate your brain. Your brain is flexible and it continues to grow throughout your life span—but only if provided the right stimuli. Studying mathematics and persistently working through tough math problems is one way to promote increased brain function. So, when doing mathematics, remember the 3 P's—Practice, Patience, and Persistence—and the positive effects they will have on your brain!

Questions

1. What is another area (not mentioned here) that requires practice, patience, and persistence to master? Can you think of anything you could master without practice?

2. Can you think of an example in your study of math where practice, patience, and persistence have helped you improve?

Strategies for Academic Success 🎓

Note Taking

Taking notes in class is an important step in understanding new material. While there are several methods for taking notes, every note-taking method can benefit from these general tips.

General Tips

- Write the date and the course name at the top of each page.
- Write the notes in your own words and paraphrase.
- Use abbreviations, such as ft for foot, # for number, def for definition, and RHS for right-hand side.
- Copy all figures or examples that are presented during the lecture.
- Review and rewrite your notes after class. Do this on the same day, if possible.

There are many different methods of note taking and it's always good to explore new methods. A good time to try out new note-taking methods is when you rewrite your class notes. Be sure to try each new method a few times before deciding which works best for you. Presented here are three note-taking methods you can try out. You may even find that a blend of several methods works best for you.

Note-Taking Methods

Outline

An outline consists of several topic headings, each followed by a series of indented bullet points that include subtopics, definitions, examples, and other details.

Example:

1. Ratio
   a. Comparison of two quantities by division.
   b. Ratio of a to b
      i. $\dfrac{a}{b}$
      ii. $a : b$
      iii. a to b
   c. Can be reduced
   d. Common units can cancel

Split Page

The split page method divides the page vertically into two columns with the left column narrower than the right column. Main topics go in the left column and detailed comments go in the right column. The bottom of the page is reserved for a short summary of the material covered.

Example:

Keywords:	Notes:
Ratios	1. Comparison of two quantities by division
	2. $\dfrac{a}{b}$, $a : b$, a to b
	3. Can reduce
	4. Common units can cancel

Summary: Ratios are used to compare quantities and units can cancel.

Mapping

The mapping method is the most visual of the three methods. One common way to create a mapping is to write the main idea or topic in the center and draw lines, from the main idea to smaller ideas or subtopics. Additional branches can be created from the subtopics until all of the key ideas and definitions are included. Using a different color for subtopic can help visually organize the topics.

Example:

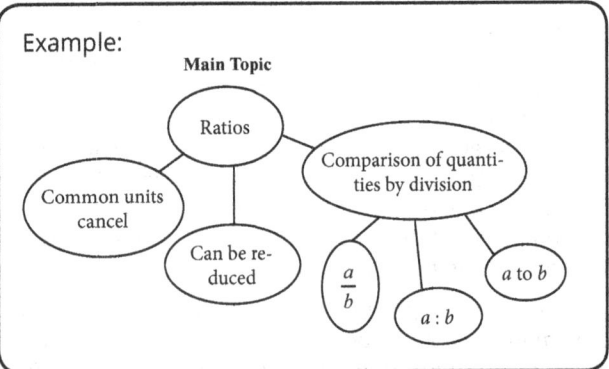

Questions

1. Find two other note taking methods and describe them.

2. Write five additional abbreviations that you could use while taking notes.

Strategies for Academic Success 🎓

Do I Need a Math Tutor?

If you do not understand the material being presented in class, if you are struggling with completing homework assignments, or if you are doing poorly on tests, then you may need to consider getting a tutor. In college, everyone needs help at some point in time. What's important is to recognize that you need help before it's too late and you end up having to retake the class.

Alternatives to Tutoring

Before getting a tutor, you might consider setting up a meeting with your instructor during their office hours to get help. Unfortunately, you may find that your instructor's office hours don't coincide with your schedule or don't provide enough time for one-on-one help.

Another alternative is to put together a study group of classmates from your math class. Working in groups and explaining your work to others can be very beneficial to your understanding of mathematics. Study groups work best if there are three to six members. Having too many people in a study group may make it difficult to schedule a time for all group members to meet. A large study group may also increase distractions. If you have too few people and those that attend are just as lost as you, then you aren't going to be helpful to each other.

Where to Find a Tutor

Many schools have both group and individual tutoring available. In most cases, the cost of this tutoring is included in tuition costs. If your college offers tutoring through a learning lab or tutoring center, then you should take advantage of it. You may need to complete an application to be considered for tutoring, so be sure to get the necessary paperwork at the start of each semester to increase your chances of getting a tutoring time that works well with your schedule. This is especially important if you know that you struggle with math or haven't taken any math classes in a while.

If you find that you need more help than the tutoring center can provide, or your school doesn't offer tutoring, you can hire a private tutor. The hourly cost to hire a private tutor varies significantly depending on the area you live in along with the education and experience level of the tutor. You might be able to find a tutor by asking your instructor for references or by asking friends who have taken higher-level math classes than you have. You can also try researching the internet for local reputable tutoring organizations in your area.

What to Look for in a Tutor

Whether you obtain a tutor through your college or hire a personal tutor, look for someone who has experience, educational qualifications, and who is friendly and easy to work with. If you find that the tutor's personality or learning style isn't similar to yours, then you should look for a different tutor that matches your style. It may take some effort to find a tutor who works well with you.

How to Prepare for a Tutoring Session

To get the most out of your tutoring session, come prepared by bringing your text, class notes, and any homework or questions you need help with. If you know ahead of time what you will be working on, communicate this to the tutor so they can also come prepared. You should attempt the homework prior to the session and write notes or questions for the tutor. Do not use the tutor to do your homework for you. The tutor will explain to you how to do the work and let you work some problems on your own while he or she observes. Ask the tutor to explain the steps aloud while working through a problem. Be sure to do the same so that the tutor can correct any mistakes in your reasoning. Take notes during your tutoring session and ask the tutor if he or she has any additional resources such as websites, videos, or handouts that may help you.

Questions

1. It's important to find a tutor whose learning style is similar to yours. What are some ways that learning styles can be different?

2. What sort of tutoring services does your school offer?

Strategies for Academic Success 🎓

Tips for Improving Your Memory

Experts believe that there are three ways that we store memories: first in the sensory stage, then in short term memory, and finally in long term memory.[1] Because we can't retain all the information that bombards us daily, the different stages of memory act as a filter. Your sensory memory lasts only a fraction of a second and holds your perception of a visual image, a sound, or a touch. The sensation then moves to your short term memory, which has the limited capacity to hold about seven items for no more than 20 to 30 seconds at a time. Important information is gradually transferred to long term memory. The more the information is repeated or used, the greater the chance that it will end up in long term memory. Unlike sensory and short term memory, long term memory can store unlimited amounts of information indefinitely. Here are some tips to improve your chances of moving important information to long-term memory.

Be attentive and focused on the information.

Study in a location that is free of distractions and avoid watching TV or listening to music with lyrics while studying.

Recite information aloud.

Ask yourself questions about the material to see if you can recall important facts and details. Pretend you are teaching or explaining the material to someone else. This will help you put the information into your own words.

Associate the information with something you already know.

Think about how you can make the information personally meaningful—how does it relate to your life, your experiences, and your current knowledge? If you can link new information to memories already stored, you create "mental hooks" that help you recall the information. For example, when trying to remember the formula for slope using rise and run, remember that rise would come alphabetically before run, so rise will be in the numerator in the slope fraction and run will be in the denominator.

Use visual images like diagrams, charts, and pictures.

You can make your own pictures and diagrams to help you recall important definitions, theorems, or concepts.

Split larger pieces of information into smaller "chunks."

This is useful when remembering strings of numbers, such as social security numbers and telephone numbers. Instead of remembering a sequence of digits such as 555777213 you can break it into chunks such as 555 777 213.

Group long lists of information into categories that make sense.

For example, instead of remembering all the properties of real numbers individually, try grouping them into shorter lists by operation, such as addition and multiplication.

Use mnemonics or memory techniques to help remember important concepts and facts.

A mnemonic that is commonly used to remember the order of operations is "Please Excuse My Dear Aunt Sally," which uses the first letter of the words Parentheses, Exponents, Multiplication, Division, Addition, and Subtraction to help you remember the correct order to perform basic arithmetic calculations. To make the mnemonic more personal and possibly more memorable, make up one of your own.

Use acronyms to help remember important concepts or procedures.

An acronym is a type of mnemonic device which is a word made up by taking the first letter from each word that you want to remember and making a new word from the letters. For example, the word HOMES is often used to remember the five Great Lakes in North America where each letter in the word represents the first letter of one of the lakes: Huron, Ontario, Michigan, Erie, and Superior.

Questions

1. Create an original mnemonic or acronym for any math topic covered so far in this course.

2. Explain two ways you can incorporate these tips into your study routine.

Strategies for Academic Success 🎓

Overcoming Anxiety

People who are anxious about math are often just not good at taking math tests. If you understand the math you are learning but don't do well on math tests, you may be in the same situation. If there are other subject areas in which you also perform poorly on tests, then you may be experiencing test anxiety.

How to Reduce Math Anxiety

- Learn effective math study skills. Sit near the front of your class and take notes. Ask questions when you don't understand the material. Review your notes after class and read new material before it's covered in class. Keep up with your assignments and do a lot of practice problems.

- Don't accept negative self talk such as "I am not good at math" or "I just don't get it and never will." Maintain a positive attitude and set small math achievement goals to keep you positively moving toward bigger goals.

- Visualize yourself doing well in math, whether it's on a quiz or test, or passing a math class. Rehearse how you will feel and perform on an upcoming math test. It may also help to visualize how you will celebrate your success after doing well on the test.

- Form a math study group. Working with others may help you feel more relaxed about math in general and you may find that other people have the same fears.

- If you panic or freeze during a math test, try to work around the panic by finding something on the math test that you can do. Once you gain confidence, work through other problems you know how to do. Then, try completing the harder problems, knowing that you have a large part of the test completed already.

- If you have trouble remembering important concepts during tests, do what is called a "brain drain" and write down all the formulas and important facts that you have studied on your test or scratch paper as soon as you are given the test. Do this before you look at any questions on the test. Having this information available to you should help boost your confidence and reduce your anxiety. Doing practice brain drains while studying can help you remember the concepts when the test time comes.

How to Reduce Test Anxiety

- Be prepared. Knowing you have prepared well will make you more confident and less anxious.

- Get plenty of sleep the night before a big test and be sure to eat nutritious meals on the day of the test. It's helpful to exercise regularly and establish a set routine for test days. For example, your routine might include eating your favorite food, putting on your lucky shirt, and packing a special treat for after the test.

- Talk to your instructor about your anxiety. Your instructor may be able to make accommodations for you when taking tests that may make you feel more relaxed, such as extra time or a more calming testing place.

- Learn how to manage your anxiety by taking deep, slow breaths and thinking about places or people who make you happy and peaceful.

- When you receive a low score on a test, take time to analyze the reasons why you performed poorly. Did you prepare enough? Did you study the right material? Did you get enough rest the night before? Resolve to change those things that may have negatively affected your performance in the past before the next test.

- Learn effective test taking strategies. See the study skill on Tips for Improving Math Test Scores.

Questions

1. Describe your routine for test days. Think of two ways you can improve your routine to reduce stress and anxiety.

2. Research and describe the accommodations that your instructor or school can provide for test taking.

Strategies for Academic Success 🎓

Online Resources

With the invention of the internet, there are numerous resources available to students who need help with mathematics. Here are some quality online resources that we recommend.

HawkesTV

tv.hawkeslearning.com

If you are looking for instructional videos on a particular topic, then start with HawkesTV. There are hundreds of videos that can be found by looking under a particular math subject area such as introductory algebra, precalculus, or statistics. You can also find videos on study skills.

YouTube

www.youtube.com

You can also find math instructional videos on YouTube, but you have to search for videos by topic or key words. You may have to use various combinations of key words to find the particular topic you are looking for. Keep in mind that the quality of the videos varies considerably depending on who produces them.

Google Hangouts

plus.google.com/hangouts

You can organize a virtual study group of up to 10 people using Google Hangouts. This is a terrific tool when schedules are hectic and it avoids everyone having to travel to a central location. You do have to set up a Google+ profile to use Hangouts. In addition to video chat, the group members can share documents using Google Docs. This is a great tool for group projects!

Wolfram | Alpha

www.wolframalpha.com

Wolfram | Alpha is a computational knowledge engine developed by Wolfram Research that answers questions posed to it by computing the answer from "curated data." Typical search engines search all of the data on the Internet based on the key words given and then provide a list of documents or web pages that might contain relevant information. The data used by Wolfram | Alpha is said to be "curated" because someone has to verify its integrity before it can be added to the database, therefore ensuring that the data is of high quality. Users can submit questions and request calculations or graphs by typing their request into a text field. Wolfram | Alpha then computes the answers and related graphics from data gathered from both academic and commercial websites such as the CIA's World Factbook, the United States Geological Survey, financial data from Dow Jones, etc. Wolfram | Alpha uses the basic features of Mathematica, which is a computational toolkit designed earlier by Wolfram Research that includes computer algebra, symbol and number computation, graphics, and statistical capabilities.

Questions

1. Describe a situation where you think Wolfram | Alpha might be more helpful than YouTube, and vice versa.

2. What are some pros and cons to using Google Hangouts?

Strategies for Academic Success 🎓

Preparing for a Final Math Exam

Since math concepts build on one another, a final exam in math is not one you can study for in a night or even a day or two. To pull all the concepts together for the semester, you should plan to start one or two weeks ahead of time. Being comfortable with the material is key to going into the exam with confidence and lowering your anxiety.

Before You Start Preparing for the Exam

1. What is the date, time, and location of the exam? Check your syllabus for the final exam time and location. If it's not on your syllabus, your instructor should announce this information in class.

2. Is there a time limit on the exam? If you experience test anxiety on timed tests, be sure to speak to your professor about it and see if you can receive accommodations that will help reduce your anxiety, such as extended time or an alternate testing location.

3. Will you be able to use a formula sheet, calculator, and/or scrap paper on the exam? If you are not allowed to use a formula sheet, you should write down important formulas and memorize them. Most of the time, math professors will advise you of the formulas you need to know for an exam. If you cannot use a calculator on the exam, be sure to practice doing calculations by hand when you are preparing for the exam and go back and check them using the calculator.

A Week Before the Exam

1. Decide where to study for the exam and with whom. Make sure it's a comfortable study environment with few outside distractions. If you are studying with others, make sure the group is small and that the people in the group are motivated to study and do well on the exam. Plan to have snacks and water with you for energy and to avoid having to delay studying to go get something to eat or drink. Be sure and take small breaks every hour or two to keep focused and minimize frustration.

2. Organize your class notes and any flash cards with vocabulary, formulas, and theorems. If you haven't used flash cards for vocabulary, go back through your notes and highlight the vocabulary. Create a formula sheet to use on the exam, if the professor allows. If not, then you can use the formula sheet to memorize the formulas that will be on the exam.

3. Start studying for the exam. Studying a week before the exam gives you time to ask your instructor questions as you go over the material. Don't spend a lot of time reviewing material you already know. Go over the most difficult material or material that you don't understand so you can ask questions about it. Be sure to review old exams and work through any questions you missed.

3 Days Before the Exam

1. Make yourself a practice test consisting of the problem types. Don't necessarily put the questions in the order that the professor covered them in class.

2. Ask your instructor or classmates any questions that you have about the practice test so that you have time to go back and review the material you are having difficulty with.

The Night Before the Exam

1. Make sure you have all the supplies you will need to take the exam: formula sheet and calculator, if allowed, scratch paper, plain and colored pencils, highlighter, erasers, graph paper, extra batteries, etc.

2. If you won't be allowed to use your formula sheet, review it to make sure you know all the formulas. Right before going to bed, review your notes and study materials, but do not stay up all night to "cram."

3. Go to bed early and get a good night's sleep. You will do better if you are rested and alert.

The Day of the Exam

1. Get up with plenty of time to get to your exam without rushing. Eat a good breakfast and don't drink too much caffeine, which can make you anxious.

2. Review your notes, flash cards, and formula sheet again, if you have time.

3. Get to class early so you can be organized and mentally prepared.

Checklist for the Exam

Date of the Exam: _____   **Time of the Exam:** _____

Location of the Exam: _____

Items to bring to the exam:

___ calculator and extra batteries   ___ pencils

___ formula sheet   ___ eraser

___ scratch paper   ___ colored pencils or highlighter

___ graph paper   ___ ruler or straightedge

Notes or other things to remember for exam day:

During the Exam

1. Put your name at the top of your exam immediately. If you are not allowed to use a formula sheet, before you even look at the exam, do what is called a "brain drain" or "data dump." Recall as much of the information on your formula sheet as you possibly can and write it either on the scratch paper or in the exam margins if scratch paper is not allowed. You have now transferred over everything on your "mental cheat sheet" to the exam to help yourself as you work through the exam.

2. Read the directions carefully as you go through the exam and make sure you have answered the questions being asked. Also, check your solutions as you go. If you do any work on scratch paper, write down the number of the problem on the paper and highlight or circle your answer. This will save you time when you review the exam. The instructor may also give you partial credit for showing your work. (Don't forget to attach your scratch work to your exam when you turn it in.)

3. Skim the questions on the exam, marking the ones you know how to do immediately. These are the problems you will do first. Also note any questions that have a higher point value. You should try to work these next or be sure to leave yourself plenty of time to do them later.

4. If you get to a problem you don't know how to do, skip it and come back after you finish all the ones you know how to do. A problem you do later may jog your memory on how to do the problem you skipped.

5. For multiple choice questions, be sure to work the problem first before looking at the answer choices. If your answer is not one of the choices, then review your math work. You can also try starting with the answer choices and working backwards to see if any of them work in the problem. If this doesn't work, see if you can eliminate any of the answer choices and make an educated guess from the remaining ones. Mark the problem to come back to later when you review the exam.

6. Once you have an answer for all the problems, review the entire exam. Try working the problems differently and comparing the results or substituting the answers into the equation to verify they are correct. Do not worry about finishing early. You are in control of your own time—and your own success!

Questions

1. Does your syllabus provide any of the information needed for the checklist?

2. Are there any tips or suggestions mentioned here that you haven't thought of before?

Strategies for Academic Success 🎓

Managing Your Time Effectively

Have you ever made it to the end of a day and wondered where all of your time went? Sometimes it feels like there aren't enough hours in the day. Managing your time is important because you can never get that time back. Once it's gone, you have to rush and cram the work into your schedule. Not only will you start feeling stressed out, but you may also find yourself turning in late or incomplete work.

Here are three strategies for managing your time more effectively.

🕐 Time Budgets

Time budgets help you find the time you need to complete necessary projects and tasks. Just like a financial budget shows you how you spend your money, a time budget shows you how you spend your time. You can then identify "wasted" time that could be used more productively.

To begin budgeting your time, assess how much time each week you spend on different types of activities, like Sleep, Meals, Work, Class, Study, Extracurricular, Exercise, Personal, Other, etc.

- What are some activities you'd like to spend more time doing in the future?

- What are some activities you should spend less time doing in the future?

Based on your answers to the questions above, create a weekly time budget. One week contains only 168 hours. If you want to spend more time on a particular activity, you'll need to find that time somewhere. Use a planner to schedule specific blocks of time for study sessions, meals, travel times, and morning/evening routines. As a general rule, you should set aside at least two hours of study time for every one hour of class time. That means that a three-credit course would require at least six hours of outside work per week.

⚖️ Breaks

When you are working on an important project or studying for a big exam, you can feel tempted to go as long as possible without taking a break. While staying focused is important, working yourself until you're mentally drained will lower the quality of your work and force you to take even more time recovering.

Just like taking breaks helps your physical body recover, it will also help your brain re-energize and refocus. During study sessions, you should plan to take a break at least once an hour. Study and work breaks should usually last around five minutes. The longer the break, the harder it is to start working again. Some courses have a built-in break during the middle of the class period. Stand up and move around, even if you don't feel tired. Even this little bit of physical movement can help you think more clearly.

📋 Avoiding Multitasking

Multitasking is working on more than one task at a time. When you have several assignments that need to be completed, you may be tempted to save time by working on two or three of them at once. While this strategy might seem like a time-saver, you will probably end up using more time than if you had done each task individually. Not only will you have to switch your focus from one task to the next, but you will also make more mistakes that will need to be corrected later. Multitasking usually ends up wasting time instead of saving it.

Instead of trying to do two things at once, schedule yourself time to work on one task at a time. To-do lists can be helpful tools for keeping yourself focused on finishing one item before moving on to another. You'll do better work and save yourself time.

Questions

1. Are there any areas in your day that are taking up too much of your time, making it hard to devote enough time to more important things?

2. Can you think of a time when multitasking has resulted in lower quality outcome in your experience?

review concepts

CHAPTER 1.R

Number Systems and Fundamental Concepts of Algebra

1.R.1 Exponents, Prime Numbers, and LCM

1.R.2 Reducing Fractions to Lowest Terms

1.R.3 Decimals and Percents

1.R.4 Simplifying Radicals

1.R.1 Exponents, Prime Numbers, and LCM

1. When looking at $3^5 = 243$, 3 is the _____, 5 is the _____, and 243 is the _____. _____ are written slightly to the right and above the _____. The expression 3^5 is _____.

The Exponent 1

Any number raised to the first power _____

For example, _____

DEFINITION

The Exponent 0

Any nonzero number raised to the 0 power _____

For example, _____

Note: The expression 0^0 _____

DEFINITION

Rules for Order of Operations

1. Simplify within grouping symbols, such as _____ (If there are more than one pair of grouping symbols, start with _____ _____

2. Evaluate any _____

3. Moving from left to right, perform any _____ _____

4. Moving from left to right, perform any _____ _____

PROCEDURE

---

2. A well-known mnemonic device for remembering the rules for order of operations is the following.

Please	Excuse	My	Dear	Aunt	Sally
↓	↓	↓	↓	↓	↓
_____	_____	_____	_____	_____	_____

▶ Watch and Work

Watch the video for Example 5 in the software and follow along in the space provided.

Example 5 Using the Order of Operations with Whole Numbers

Simplify: $2 \cdot 3^2 + 18 \div 3^2$

Solution

✐ Now You Try It!

Use the space provided to work out the solution to the next example.

Example A Using the Order of Operations with Whole Numbers

Simplify: $6^2 \div 9 + 3 - 14 \div 7$

Divisibility

If a number can be divided by another number so that the remainder is 0, then we say

1. the number is _____

2. the divisor _____

DEFINITION

Divisibility by 2

A number is divisible by 2 (is an **even number**) if _____

DEFINITION

Divisibility by 3

A number is divisible by 3 if _____

DEFINITION

▶ Watch and Work

Watch the video for Example 8 in the software and follow along in the space provided.

Example 8 Determining Divisibility by 3

Determine whether each of the following numbers is divisible by 3.

a. 6801

b. 356

Solution

✏️ Now You Try It!

Use the space provided to work out the solution to the next example.

Example A Determining Divisibility by 3

Is 7912 divisible by 3? Explain why or why not.

Divisibility by 4

A number is divisible by 4 if _____
_____

DEFINITION

Divisibility by 5

A number is divisible by 5 if _____

DEFINITION

Divisibility by 6

A number is divisible by 6 if _____

DEFINITION

Divisibility by 9

A number is divisible by 9 if _____

DEFINITION

Divisibility by 10

A number is divisible by 10 if _____

DEFINITION

Prime Number

A **prime number** is a counting number _____

_____

DEFINITION

Composite Number

A **composite number** is _____

DEFINITION

To Determine Whether a Number is Prime

Divide the number by progressively larger prime numbers (2, 3, 5, 7, 11, and so forth) until one of the following is true.

3. The remainder _____ This means that the _____

_____

4. You find a quotient _____ This means that the _____

_____

PROCEDURE

The Fundamental Theorem of Arithmetic

Every composite number has _____

DEFINITION

To Find the Prime Factorization of a Composite Number

1. Factor the composite number _____

2. Factor each _____

3. Continue this process until all factors are prime.

PROCEDURE

Factors of a Composite Number

The only factors (or divisors) of a composite number are

1. _____

2. _____

3. products formed by _____

DEFINITION

▶ Watch and Work

Watch the video for Example 22 in the software and follow along in the space provided.

Example 22 Finding the Factors of a Composite Number

Find all the factors of 60.

Solution

✏ Now You Try It!

Use the space provided to work out the solution to the next example.

Example A Finding the Factors of a Composite Number

Find all the factors of 42.

The **multiples** of a number are _____

> ## Least Common Multiple (LCM)
>
> The **least common multiple (LCM)** of two (or more) counting numbers is _____
>
> _____
>
> **DEFINITION**

> ## To Find the LCM of a Set of Counting Numbers
>
> 1. Find the _____
> 2. List the _____
> 3. Find the product of these primes using each _____
>   _____
>
> **PROCEDURE**

▶ Watch and Work

Watch the video for Example 27 in the software and follow along in the space provided.

Example 27   Finding the Least Common Multiple (LCM)

Find the LCM of 27, 30, and 42.

Solution

✏ Now You Try It!

Use the space provided to work out the solution to the next example.

Example A Finding the Least Common Multiple (LCM)

Find the LCM of 36, 45, and 60.

Finding Equivalent Fractions

To find a fraction equivalent to $\frac{a}{b}$, multiply the _____

_____

$$\frac{a}{b} = \frac{a}{b} \cdot \underline{\hspace{3cm}}$$

For example, _____

PROCEDURE

1.R.1 Exercises

Concept Check

True/False. Determine whether each statement is true or false. If a statement is false, explain how it can be changed so the statement will be true. (**Note:** There may be more than one acceptable change.)

1. Nine squared is equal to eighteen.

2. $2^7 = 128$

3. 7^0 is undefined.

4. According to the order of operations, multiplication is always performed before division.

5. A number that is divisible by 10 is also divisible by 2 and 5.

6. 6801 is divisible by 9.

7. 7605 is divisible by 10.

8. 5,187,042 is divisible by 3.

9. A prime number has exactly 1 factor.

10. A composite number has 2 or more factors.

11. 231 is a prime number.

12. All the factors of 30 are 1, 2, 3, 5, 6, 10, 15 and 30.

13. The LCM of 15 and 25 is 50.

14. The first five multiples of 9 are 9, 18, 27, 36, and 45.

15. When given larger numbers, the most efficient way to find the LCM is to use the prime factorization method.

Practice

For each exponential expression **a.** identify the base, **b.** identify the exponent, and **c.** evaluate the exponential expression.

16. 2^3 **17.** 4^0

Simplify.

18. $30 \div 2 - 11 + 2(5-1)^3$

Using the tests for divisibility, determine which of 2, 3, 4, 5, 6, 9, and 10 (if any) will divide exactly into each given number.

19. 105

21. 331

22. 1234

20. 150

Determine whether each number is prime or composite. If the number is composite, find at least three factors of the number.

23. 47

24. 63

Find the prime factorization of each number. Use the tests for divisibility for 2, 3, 4, 5, 6, 9, and 10 whenever they help to find beginning factors.

25. 125

26. 150

27. Find the LCM of 3, 4, and 8.

28. For 14, 35, and 49, **a.** find the LCM and **b.** state how many times each number divides into the LCM.

For each equation, find the missing numerator that will make the fractions equivalent.

29. $\dfrac{5}{8} = \dfrac{?}{24}$

30. $\dfrac{5}{12} = \dfrac{?}{108}$

Applications

Solve.

31. **Purchases:** Robert is purchasing shirts for his weekend soccer team. The shirts he wants to buy are normally $25 each but are on sale for $10 off. His team has a total of 11 players. How much will he spend to buy the shirts?

   a. If you simplify the expression $25 − $10 · 11 using the order of operations, will you get the correct answer? If not, explain what is wrong with the expression.

   b. What is the answer? If necessary, write the corrected expression to get the correct results when following the order of operations.

32. **Fundraising:** You are on a team that is participating in a charity walk with a goal to raise $12,400. Each team member agrees to raise the same amount of money. If the possible team sizes are 5, 6, 9, or 10 members, which team sizes allow the goal amount to be evenly split between the team members? How much money would each team member raise for each team size that can evenly split the goal amount?

33. **Time:** A company is working on a project that will take 440 hours of work to complete. The manager in charge of the project has the option to have 4, 6, or 8 people work on the project. If the manager wants to evenly divide the work between the team members, which team size will evenly split the work hours? How many hours would each team member spend on the project for each team size that evenly splits the work hours?

34. **Inventory:** Twenty-four pencils are to be distributed evenly between the members of a group. What are the possible group sizes if each person in the group is to receive the same number of pencils?

35. *Baking:* A chocolatier makes 72 specialty truffles. She wants to sell packages that each have the same number of truffles. What are her options for the number of truffles that can be in a package?

36. *Security:* Three security guards meet at the front gate for coffee before they walk around inspecting buildings at a manufacturing plant. The guards take 15, 20, and 30 minutes, respectively, for the inspection trip.

    a. If they start at the same time, in how many minutes will they meet again at the front gate for coffee?

    b. How many trips will each guard have made?

Writing & Thinking

37. Give one example where addition should be completed before multiplication.

38. a. If a number is divisible by both 3 and 5, then it will be divisible by 15. Give two examples.

    b. However, a number might be divisible by 3 and not by 5. Give two examples.

    c. Also, a number might be divisible by 5 and not 3. Give two examples.

39. Are all odd numbers also prime numbers? Explain your answer.

40. Explain the difference between factors of a number and multiples of that number.

41. Explain, in your own words, why each number in a set divides evenly into the LCM of that set of numbers.

42. Explain why simply multiplying two numbers together will not necessarily find the LCM of those numbers. Give an example of when it would find the LCM and an example when it would not.

1.R.2 Reducing Fractions to Lowest Terms

1. A fraction is reduced to lowest terms if the numerator and denominator _____

_____

To Reduce a Fraction to Lowest Terms

1. Factor the _____

2. Use the fact that _____

Note: Reduced fractions may be improper fractions.

PROCEDURE

▶ Watch and Work

Watch the video for Example 7 in the software and follow along in the space provided.

Example 7 Multiplying and Reducing Using Prime Factors

Multiply and reduce to lowest terms: $\dfrac{17}{50} \cdot \dfrac{25}{34} \cdot 8$

Solution

✏️ Now You Try It!

Use the space provided to work out the solution to the next example.

Example A Multiplying and Reducing Using Prime Factors

Multiply and reduce to lowest terms:

$$16 \cdot \frac{12}{100} \cdot \frac{5}{36}$$

1.R.2 Exercises

Concept Check

True/False. Determine whether each statement is true or false. If a statement is false, explain how it can be changed so the statement will be true. (**Note:** There may be more than one acceptable change.)

1. $\frac{1}{2}$ and $\frac{9}{18}$ are equivalent fractions.

2. The number 1 is always a factor of the numerator and the denominator.

Practice

Multiply and reduce to lowest terms. (**Hint:** Factor before multiplying.)

3. $\dfrac{1}{3} \cdot \dfrac{3}{4}$

4. $\dfrac{2}{3} \cdot \dfrac{4}{3}$

5. $\dfrac{5}{16} \cdot \dfrac{16}{15}$

6. $\dfrac{9}{10} \cdot \dfrac{35}{40} \cdot \dfrac{25}{15}$

Applications

Solve.

7. ***Demographics:*** A study showed that $\frac{3}{5}$ of the students in an elementary school were left-handed. If the school had an enrollment of 600 students, how many were left-handed?

Writing & Thinking

8. If two fractions are between 0 and 1, can their product be more than 1? Explain.

1.R.3 Decimals and Percents

The word percent comes from the Latin *per centum*, meaning _____. So **percent means** _____, or **the ratio of a number to** _____.

The symbol % is called the _____. This sign has the same meaning as the fraction $\frac{1}{100}$

To Change a Decimal Number to a Percent

1. Move the _____

2. Write the _____

PROCEDURE

To Change a Percent to a Decimal Number

1. Move the _____

2. Delete the _____

PROCEDURE

▶ Watch and Work

Watch the video for Example 3 in the software and follow along in the space provided.

Example 3 Changing Percents to Decimal Numbers

Change each percent to a decimal number.

a. 76.%

b. 18.5%

c. 50%

d. 100%

e. 0.25%

✏ Now You Try It!

Use the space provided to work out the solution to the next example.

Example A Changing Percents to Decimal Numbers

Change each percent to a decimal number.

a. 40%

d. 29.37%

b. 211%

e. 102%

c. 0.6%

To Change a Fraction to a Percent

1. Change the _____. (Divide _____.)

2. Change the _____.

PROCEDURE

▶ Watch and Work

Watch the video for Example 6 in the software and follow along in the space provided.

Example 6 Changing Mixed Numbers to Percents

Change $2\frac{1}{4}$ to a percent.

Solution

✏ Now You Try It!

Use the space provided to work out the solution to the next example.

Example A Changing Mixed Numbers to Percents

Change $1\frac{1}{2}$ to a percent.

To Change a Percent to a Fraction or a Mixed Number

1. Write the percent as a fraction with _____

2. Reduce the _____

PROCEDURE

1.R.3 **Exercises**

Concept Check

True/False. Determine whether each statement is true or false. If a statement is false, explain how it can be changed so the statement will be true. (**Note:** There may be more than one acceptable change.)

1. If a decimal number is less than 1, then the equivalent percent will be less than 100%.

2. It is not possible to have a percent greater than 100%.

3. A decimal number that is between 0.01 and 0.10 is between 10% and 100%.

4. To change from a percent to a decimal, simply omit the percent sign.

5. Fractions that have denominators other than 100 cannot be changed to a percent.

6. The fraction $\frac{1}{5}$ is equivalent to $\frac{1}{5}$%.

Practice

Change each fraction to a percent.

7. $\frac{20}{100}$

8. $\frac{125}{100}$

Change each decimal number to a percent.

9. 0.02

10. 2.3

Change each percent to a decimal number.

11. 7%

12. 179%

Change each fraction or mixed number to a percent. If necessary, round to the nearest tenth of a percent.

13. $\frac{3}{4}$

14. $5\frac{3}{10}$

Change each percent to a fraction or mixed number and reduce, if possible.

15. 120%

16. 12.5%

Applications

Solve.

17. *Interest:* A savings account is offering an interest rate of 0.04 for the first year after opening the account. Change 0.04 to a percent.

18. *Sales Tax:* Suppose that sales tax is figured at 7.25%. Change 7.25% to a decimal.

19. *Exam Grades:* Out of a possible total of 240 points on an exam, David received 204 points. What percent of the exam did David get correct?

20. *College Degrees:* To receive a Bachelor of Science (BS) degree at Bluefield State College, the student must complete a total of 128 credit hours, of which 41 of these credits must be general education Core Skills courses. What percent of the total curriculum is dedicated to general education courses? [1]

Writing & Thinking

21. Describe the relationship between percent and the number 100.

22. Describe a situation where more than 100% is possible. Describe a situation where it is impossible to have more than 100%.

---

1   Source: 2010–2011 Bluefield, WV State College Catalogue, p.76

23. Justify why mixed numbers are a larger percentage than proper fractions alone. (Consider the value of 100%.)

24. Describe the process to change a percent to a fraction or mixed number.

1.R.4 Simplifying Radicals

Properties of Square Roots

If a and b are **positive** real numbers, then

1. _____

2.

   _____

PROPERTIES

Simplest Form for Square Roots

A square root is considered to be in **simplest form** when _____

_____

DEFINITION

Square Root of x^2

If x is a real number, then _____

Note: If $x \geq 0$ is given, then _____

DEFINITION

▶ Watch and Work

Watch the video for Example 3 in the software and follow along in the space provided.

Example 3 Simplifying Square Roots with Variables

Simplify each of the following square roots. Look for perfect square factors and even powers of the variables. Assume that all variables represent positive real numbers.

a. $\sqrt{81x^4}$

b. $\sqrt{64x^5y}$

c. $\sqrt{18a^4b^6}$

d. $\sqrt{\dfrac{9a^{13}}{b^4}}$

Solution

✏️ **Now You Try It!**

Use the space provided to work out the solution to the next example.

Example A Simplifying Square Roots with Variables

Simplify each of the following square roots. Assume that all variables represent positive real numbers.

a. $\sqrt{16x^8}$

b. $\sqrt{100x^3y^3}$

c. $\sqrt{12x^8y^{12}}$

d. $\sqrt{\dfrac{25z^{18}}{y^8}}$

Simplest Form for Cube Roots

A cube root is considered to be in _____

DEFINITION

1.R.4 **Exercises**

Concept Check

True/False. Determine whether each statement is true or false. If a statement is false, explain how it can be changed so the statement will be true. (**Note:** There may be more than one acceptable change.)

1. Any variable term with an exponent of 5 has a perfect cube factor within that variable term.

2. The simplest form of a radical expression can be found by using prime factorization.

3. If x is a real number, then $\sqrt{x^2} = x$.

4. The term $7b\sqrt[3]{6c^2}$ is in simplified form.

Practice

Simplify each of the following radical expressions. Assume that all variables represent positive real numbers.

5. $\sqrt{162}$

6. $\sqrt{\dfrac{32}{49}}$

7. $\sqrt{24x^{11}y^2}$

8. $\sqrt[3]{56}$

9. $\sqrt[3]{-8x^8}$

Applications

Use the following two formulas associated with electricity

$I = \sqrt{\dfrac{P}{R}}$

$E = \sqrt{PR}$

P = power (in watts)
I = current (in amperes)
E = voltage (in volts)
R = resistance (in ohms, Ω)

10. *Electricity:* What is the current in amperes of a light bulb that produces 150 watts of power and has a 25 Ω resistance?

11. *Electricity:* If a light bulb has a resistance of 30 Ω and produces 90 watts of power, what is its current in amperes?

Writing & Thinking

12. Under what conditions is the expression $\sqrt{a}$ not a real number?

13. Explain why the expression $\sqrt[3]{y}$ is a real number regardless of whether $y > 0$, $y < 0$, or $y = 0$.

review concepts

CHAPTER 2.R

Equations and Inequalities of One Variable

2.R.1 Multiplication and Division with Fractions

2.R.2 Addition and Subtraction with Fractions

2.R.3 Applications: Number Problems and Consecutive Integers

2.R.4 Proportions

2.R.1 Multiplication and Division with Fractions

To Multiply Fractions

1. _____

2. _____

$$\frac{a}{b} \cdot \frac{c}{d} = \underline{\qquad} \quad (b, d \neq \underline{\quad})$$

For example,

_____

PROCEDURE

Commutative Property of Multiplication

The order of the fractions being multiplied can be _____

$$\frac{a}{b} \cdot \frac{c}{d} = \underline{\qquad} \quad (b, d \neq \underline{\quad})$$

For example, _____

PROPERTIES

Associative Property of Multiplication

The grouping of the fractions being multiplied can be _____

$$\left(\frac{a}{b} \cdot \frac{c}{d}\right) \cdot \frac{e}{f} = \underline{\qquad} \quad (b, d, f \neq \underline{\quad})$$

For example,

_____

PROPERTIES

1. A fraction is reduced to lowest terms if the numerator and denominator _____

_____

To Reduce a Fraction to Lowest Terms

1. Factor the _____

2. Use the fact that _____

Note: Reduced fractions may be improper fractions.

PROCEDURE

▶ Watch and Work

Watch the video for Example 11 in the software and follow along in the space provided.

Example 11   Multiplying and Reducing Using Prime Factors

Multiply and reduce to lowest terms: $\dfrac{17}{50} \cdot \dfrac{25}{34} \cdot 8$

Solution

Now You Try It!

Use the space provided to work out the solution to the next example.

Example A Multiplying and Reducing Using Prime Factors

Multiply and reduce to lowest terms:

$16 \cdot \dfrac{12}{100} \cdot \dfrac{5}{36}$

Reciprocals

The reciprocal of $\dfrac{a}{b}$ is _____ . The product of a nonzero number and its reciprocal is

_____

$$\frac{a}{b} \cdot \frac{b}{a} = 1$$

Note: $0 = \dfrac{0}{1}$, but $\dfrac{1}{0}$ _____

DEFINITION

To Divide Fractions

To divide by any nonzero number, _____

$$\frac{a}{b} \div \frac{c}{d} = $$ _____

For example, _____

PROCEDURE

▶ Watch and Work

Watch the video for Example 22 in the software and follow along in the space provided.

Example 22 Dividing and Reducing Fractions

Divide and reduce to lowest terms: $\dfrac{16}{27} \div \dfrac{8}{9}$

Solution

✎ Now You Try It!

Use the space provided to work out the solution to the next example.

Example A Dividing and Reducing Fractions

Divide and reduce to lowest terms: $\dfrac{10}{21} \div \dfrac{5}{14}$

2.R.1 Exercises

Concept Check

True/False. Determine whether each statement is true or false. If a statement is false, explain how it can be changed so the statement will be true. (**Note:** There may be more than one acceptable change.)

1. To find $\frac{1}{2}$ of $\frac{2}{9}$ requires multiplication.

2. $\frac{3}{4} \cdot \frac{9}{10} = \frac{27}{40}$

3. The statement $\frac{1}{3} \cdot \frac{2}{5} = \frac{2}{5} \cdot \frac{1}{3}$ is an example of the associative property of multiplication.

4. The number 1 is always a factor of the numerator and the denominator.

5. The reciprocal of 1 is undefined.

6. The product of a nonzero number and its reciprocal is undefined.

7. The reciprocal of 12 is $\frac{12}{1}$.

8. The result of $\frac{1}{3} \div \frac{1}{6}$ is 2.

Practice

Multiply and reduce to lowest terms. (**Hint:** Factor before multiplying.)

9. $\dfrac{1}{3} \cdot \dfrac{3}{4}$

11. $\dfrac{5}{16} \cdot \dfrac{16}{15}$

10. $\dfrac{2}{3} \cdot \dfrac{4}{3}$

12. $\dfrac{9}{10} \cdot \dfrac{35}{40} \cdot \dfrac{25}{15}$

Divide and reduce to lowest terms.

13. $\dfrac{2}{3} \div \dfrac{3}{4}$

15. $\dfrac{5}{6} \div 0$

14. $0 \div \dfrac{5}{6}$

16. $\dfrac{14}{15} \div \dfrac{21}{25}$

Applications

Solve.

17. *Recipes:* A recipe calls for $\frac{3}{4}$ cups of flour. How much flour should be used if only half of the recipe is to be made?

18. *Demographics:* A study showed that $\frac{3}{5}$ of the students in an elementary school were left-handed. If the school had an enrollment of 600 students, how many were left-handed?

19. *Geology:* The floor of the Atlantic Ocean is spreading apart at an average rate of $\frac{3}{50}$ of a meter per year. How long will it take for the ocean floor to spread 12 meters?

20. ***Airplane Capacity:*** An airplane is carrying 180 passengers. This is $\frac{9}{10}$ of the capacity of the airplane.

    a.   Is the capacity of the airplane more or less than 180?

    b.   If you were to multiply 180 times $\frac{9}{10}$, would the product be more or less than 180?

    c.   What is the capacity of the airplane?

Writing & Thinking

21. If two fractions are between 0 and 1, can their product be more than 1? Explain.

22. Explain the process of multiplying two fractions. Give an example of a product that cannot be reduced.

23. Explain why the number 0 has no reciprocal.

24. Is division a commutative operation? Explain briefly and give three examples using fractions to help justify your answer.

2.R.2 Addition and Subtraction with Fractions

To Add Fractions with the Same Denominator

1. _____

2. _____ $\dfrac{a}{b} + \dfrac{c}{b} =$ _____

3. _____

For example,

_____

PROCEDURE

To Add Fractions with Different Denominators

1. Find the _____

2. Change each fraction into _____

3. _____

4. _____

PROCEDURE

Commutative Property of Addition

The order of the fractions being added can be _____

$$\dfrac{a}{b} + \dfrac{c}{d} = $$ _____

For example,

_____

PROPERTIES

Associative Property of Addition

The grouping of the fractions being added can be _____

$$\dfrac{a}{b} + \left(\dfrac{c}{d} + \dfrac{e}{f} \right) = $$ _____

For example,

_____

PROPERTIES

To Subtract Fractions with the Same Denominator

1. _____

2. _____ $\dfrac{a}{b} - \dfrac{c}{b} =$ _____

3. _____

For example, _____

PROCEDURE

To Subtract Fractions with Different Denominators

1. Find the _____

2. Change each fraction into _____

3. _____

4. _____

PROCEDURE

▶ Watch and Work

Watch the video for Example 11 in the software and follow along in the space provided.

Example 11 Subtracting Fractions with Different Denominators

Subtract: $1 - \dfrac{5}{8}$

Solution

✏ Now You Try It!

Use the space provided to work out the solution to the next example.

Example A Subtracting Fractions with Different Denominators

Subtract: $3 - \dfrac{5}{12}$

2.R.2 **Exercises**

Concept Check

True/False. Determine whether each statement is true or false. If a statement is false, explain how it can be changed so the statement will be true. (**Note:** There may be more than one acceptable change.)

1. The final step in adding fractions is to reduce, if possible.

2. The process for finding the LCD is the same as the process for finding the LCM.

3. LCD represents the Least Common Digit.

4. When subtracting fractions, simply subtract the numerators and the denominators.

5. Subtraction of fractions requires that the fractions have the same denominators.

Practice

Add and reduce to lowest terms.

6. $\dfrac{3}{25} + \dfrac{12}{25}$

7. $\dfrac{2}{7} + \dfrac{4}{21} + \dfrac{1}{3}$

Subtract and reduce to lowest terms.

8. $\dfrac{7}{8} - \dfrac{5}{8}$

10. $2 - \dfrac{9}{16}$

9. $\dfrac{9}{14} - \dfrac{2}{21}$

Applications

Solve.

11. **Postage:** Three pieces of mail weigh $\frac{1}{2}$ ounce, $\frac{1}{5}$ ounce, and $\frac{3}{10}$ ounce. What is the total weight of the letters?

12. **Cooking:** A recipe calls for the following spices: $\frac{1}{2}$ teaspoon of turmeric, $\frac{1}{4}$ teaspoon of ginger, and $\frac{1}{8}$ teaspoon of cumin. What is the total quantity of these three spices?

Writing & Thinking

13. Explain how finding the LCM relates to LCDs.

14. Give an example of a situation where you might add or subtract fractions (other than in class).

2.R.3 Applications: Number Problems and Consecutive Integers

Consecutive Integers

Integers are **consecutive** if each is _____

Three consecutive integers can be represented as

_____

where n is an integer.

DEFINITION

Consecutive Even Integers

Even integers are **consecutive** if each is _____

Three consecutive even integers can be represented as

_____

where n is an **even** integer.

DEFINITION

Consecutive Odd Integers

Odd integers are **consecutive** if each is _____

Three consecutive odd integers can be represented as

_____

where n is an **odd** integer.

DEFINITION

▶ Watch and Work

Watch the video for Example 6 in the software and follow along in the space provided.

Example 6 Application: Calculating Living Expenses

Joe wants to budget $\frac{2}{5}$ of his monthly income for rent. He found an apartment he likes for $800 a month. What monthly income does he need to be able to afford this apartment?

Solution

✏ Now You Try It!

Use the space provided to work out the solution to the next example.

Example A Application: Calculating Living Expenses

Jim plans to budget $\frac{3}{7}$ of his monthly income to send his son Taylor to private school. If the school he'd like Taylor to attend costs $1200 a month, what monthly income does Jim need to be able to afford the school?

2.R.3 **Exercises**

Concept Check

True/False. Determine whether each statement is true or false. If a statement is false, explain how it can be changed so the statement will be true. (**Note:** There may be more than one acceptable change.)

1. If an odd integer is divided by 2, the remainder will be 1.

2. To find 3 consecutive odd integers, you could use n, $n + 1$, and $n + 3$.

3. Odd integers are integers that are divisible by 1.

4. Even integers are consecutive if each is 2 more than the previous even integer.

Practice

Read each problem carefully, translate the various phrases into algebraic expressions, set up an equation, and solve the equation.

5. Five less than a number is equal to 13 decreased by the number. Find the number.

6. Twice a number increased by 3 times the number is equal to 4 times the sum of the number and 3. Find the number.

7. Find three consecutive integers whose sum is 93.

8. Find three consecutive odd integers such that the sum of twice the first and three times the second is 7 more than twice the third.

Applications

Solve.

9. *Expenses:* A collect call from a landline in Ohio to another landline in Ohio has a connection fee of $2.75 and a charge of $0.36 per minute. Mr. Anderson made a collect call which cost the receiver of the call $9.95. This situation can be modeled by $9.95 = \$2.75 + \$0.36m$.

   a. The unknown value is represented by the variable m in the equation. What is the unknown value in this situation?

   b. Solve the equation for the variable.

   c. What does the answer to Part **b.** mean? Write a complete sentence.

10. *Event Planning:* Robin is in charge of purchasing desserts for a dinner party that her nonprofit organization is throwing. She decides to buy a cake and several specialty cupcakes from Barbara's Bombtastic Bakery. She needs to buy one 8-inch round cake which costs $19.50. She has $45 to spend and will spend the leftover amount on cupcakes, which are $8.50 for a box of 4. How many boxes of cupcakes can Robin purchase?

   a. What is the unknown value in this problem? Let the variable c represent this unknown value.

   b. Write an equation to represent this situation.

   c. Solve the equation for the variable.

   d. What does the answer to Part c. mean? Write a complete sentence.

Writing & Thinking

11. a. How would you represent four consecutive odd integers?

    b. How would you represent four consecutive even integers?

    c. Are these representations the same? Explain.

2.R.4 Proportions

Proportions

A **proportion** is a statement that _____

In symbols,

_____

A proportion is true if _____

DEFINITION

To Solve a Proportion

1. Find the cross products (or cross multiply) and then _____

2. Divide both sides of the equation by _____

3. _____

PROCEDURE

▶ Watch and Work

Watch the video for Example 3 in the software and follow along in the space provided.

Example 3 Solving Proportions

Find the value of x if $\dfrac{4}{8} = \dfrac{5}{x}$.

Solution

✏ Now You Try It!

Use the space provided to work out the solution to the next example.

Example A Solving Proportions

Find the value of x if $\dfrac{12}{x} = \dfrac{9}{15}$.

To Solve an Application Using a Proportion

1. Identify the unknown quantity and _____

2. Set up a proportion in which the _____. (Make sure that the _____)

3. Solve the _____

PROCEDURE

2.R.4 Exercises

Concept Check

True/False. Determine whether each statement is true or false. If a statement is false, explain how it can be changed so the statement will be true. (**Note:** There may be more than one acceptable change.)

1. A proportion is a statement that two ratios are being multiplied.

2. Cross canceling is used to determine if a proportion is true.

3. In order to solve the proportion $\dfrac{16}{36.8} = \dfrac{x}{27.6}$ we construct the equation $36.8x = 441.6$.

4. When using proportions to solve a word problem, there is only one correct way to set up the proportion.

5. The proportions $\dfrac{36 \text{ tickets}}{\$540} = \dfrac{x \text{ tickets}}{\$75}$ and $\dfrac{x \text{ tickets}}{36 \text{ tickets}} = \dfrac{\$75}{\$540}$ will yield the same answer.

Practice

Determine whether each proportion is true or false.

6. $\dfrac{3}{6} = \dfrac{4}{8}$

7. $\dfrac{1}{3} = \dfrac{33}{100}$

Solve each proportion.

8. $\dfrac{5}{4} = \dfrac{x}{8}$

9. $\dfrac{3.5}{2.6} = \dfrac{10.5}{B}$

Applications

Solve.

10. *Concrete:* The quality of concrete is based on the ratio of bags of cement to cubic yards of gravel. One batch of concrete consists of 27 bags of cement mixed into 9 cubic yards of gravel, while a second has 15 bags of cement mixed with 5 cubic yards of gravel. Determine whether the ratio of cement to gravel is the same for both batches.

11. *Grading:* An English teacher must read and grade 27 essays. If the teacher takes 20 minutes to read and grade 3 essays, how much time will he need to grade all 27 essays?

Writing & Thinking

12. In your own words, clarify how you can know that a proportion is set up correctly or not.

CHAPTER 4.R

Relations, Functions, and Their Graphs

4.R.1 Order of Operations with Real Numbers

4.R.2 Identifying Like Terms

4.R.3 Simplifying Expressions

4.R.4 Translating English Phrases and Algebraic Expressions

4.R.1 Order of Operations with Real Numbers

Rules for Order of Operations

1. Simplify within grouping symbols, such as_____
   _____

2. Find any _____

3. Moving from left to right, perform any _____

4. Moving from left to right, perform any _____

PROCEDURE

▶ Watch and Work

Watch the video for Example 3 in the software and follow along in the space provided.

Example 3 Using the Order of Operations with Real Numbers

Simplify: $(2-5)^2 + \left|2-5^2\right| - 2^3$

Solution

✎ Now You Try It!

Use the space provided to work out the solution to the next example.

Example A Using the Order of Operations with Real Numbers

Simplify: $(1-3)^2 + \left|9-4^2\right| - 1^3$

4.R.1 Exercises

True/False. Determine whether each statement is true or false. If a statement is false, explain how it can be changed so the statement will be true. (**Note:** There may be more than one acceptable change.)

1. If there are no grouping symbols, multiplication should always be performed before addition.

2. When following the rules for order of operations, powers indicated by exponents should be evaluated last.

3. The square root symbol is a grouping symbol.

4. A well-known mnemonic device for remembering the rules for order of operations is SADMEP.

Practice

Simplify.

5. **a.** $24 \div 4 \cdot 6$

   b. $24 \cdot 4 \div 6$

7. $3^2 \div (-9) \cdot (4 - 2^2) + 5(-2)$

8. $14 \cdot 3 \div (-2) - 6(4)$

6. $15 \div (-3) \cdot 3 - 10$

9. $|16 - 20| + (-10)^2 + 5^2$

Applications

Solve.

10. *Discounts:* The Matthews family, a family of 4, is planning a trip to New York City. During their visit, they want to see the Broadway play *Matilda*. The tickets cost $102 each. The Matthews purchase the tickets online and the website charges a service fee of $7.50 per ticket. The website is running a sale where the Matthews can get 10% off of their entire purchase.

   a.   Write an expression to describe how much of a discount the Matthews will receive on their purchase.

   b.   What is the final purchase price of the tickets?

11. *Banking:* Dennis overdrew his checking account and ended up with a balance of −$42. The bank charged a $35 overdraft fee and an additional $5 fee for every day the account was overdrawn. Dennis left his account overdrawn for 3 days.

   a.   Write an expression to show the balance of Dennis's checking account after 3 days.

   b.   Simplify the expression in Part **a.** to find the balance of Dennis's checking account after 3 days.

Writing & Thinking

12. Explain, in your own words, why the following expression cannot be evaluated.

$$\left(24 - 2^4\right) + 6\left(3 - 5\right) \div \left(3^2 - 9\right)$$

13. Consider any number between 0 and 1. If you square this number, will the result be larger or smaller than the original number? Is this always the case? Explain.

4.R.2 Identifying Like Terms

Like Terms

Like terms (or similar terms) are terms that are _____

_____

DEFINITION

▶ Watch and Work

Watch the video for Example 1 in the software and follow along in the space provided.

Example 1 Identifying Like Terms

Identify the like terms in the following list of terms.

$$6, \quad 2.2x, \quad -10.1, \quad -x, \quad 3x^2z, \quad -5x, \quad \frac{3}{4}x^2z, \quad 0$$

Solution

✐ Now You Try It!

Use the space provided to work out the solution to the next example.

Example A Identify the like terms in the following list of terms.

$\frac{5}{3}, 10x, -1, 4x^2z, -3x, 2x^2z, -5.65x, 0$

4.R.2 Exercises

True/False. Determine whether each statement is true or false. If a statement is false, explain how it can be changed so the statement will be true. (**Note:** There may be more than one acceptable change.)

1. A variable that does not appear to have an exponent has an exponent of 1.

2. In the term $-9x$, nine is being subtracted from x.

3. In the term "$12a$," 12 is the constant.

4. Like terms have the same coefficients.

Practice

Identify the like terms in each list of terms.

5. $-5, 3, 7x, 8, 9x, 3y$

Writing & Thinking

6. Discuss like and unlike terms and give an example of each.

4.R.3 Simplifying Expressions

Combining Like Terms

To combine like terms, _____

DEFINITION

▶ Watch and Work

Watch the video for Example 1 in the software and follow along in the space provided.

Example 1 Combining Like Terms

Simplify each expression by combining like terms.

a. $8x + 10x$

b. $6.5y - 2.3y + 3$

c. $2x^2 + 3x + x^2 - x$

d. $4(n-7) + 5(n+1)$

e. $3a - 52a^2 + 9$

f. $\dfrac{x + 3x}{2} + 5x$

Solution

✏ Now You Try It!

Use the space provided to work out the solution to the next example.

Example A Identifying the Properties of Addition and Multiplication

Simplify each expression by combining like terms.

   a. $3x + 9x$

   b. $4.2z + 5 - 3.1z$

   c. $4x^2 + 7y + 3x^2 - 2y$

   d. $6(a - 2) + 4(a + 5)$

   e. $-2x^2 + 5y^2 + 9 - y$

   f. $\dfrac{x + 5x}{3} + 6x$

4.R.3 **Exercises**

True/False. Determine whether each statement is true or false. If a statement is false, explain how it can be changed so the statement will be true. (**Note:** There may be more than one acceptable change.)

1. A variable that does not appear to have an exponent has an exponent of 1.

2. Like terms have the same coefficients.

Practice

Simplify each expression by combining like terms.

3. $8x + 7x$

4. $3x - 5x + 12x$

5. $13x + 12x^2 + 15x - 35 - 41 - 2x^2$

6. $5y + 4 - 2y$

7. $\dfrac{3a + 5a}{-2} + 12a$

Applications

Solve.

8. *Inventory:* A moving company starts the week with 72 bundles of small boxes and 50 bundles of medium boxes. During the week, they use 25 bundles of small boxes and 32 bundles of medium boxes. At the end of the week, they buy 125 bundles of medium boxes. The total number of boxes at the end of the week can be modeled by the expression $72s + 50m - 25s - 32m + 125m$, where s represents the number of boxes in a bundle of small boxes and m represents the number of boxes in a bundle of medium boxes. Simplify the expression for the total number of boxes at the end of the week by combining like terms.

4.R.4 Translating English Phrases and Algebraic Expressions

Key Words To Look For When Translating Phrases

Addition	Subtraction	Multiplication	Division	Exponent (Powers)

1. Division and subtraction are done with the values in the _____ that they are given in the problem.

2. An **ambiguous phrase** is one whose meaning is _____

   _____

▶ Watch and Work

Watch the video for Example 3 in the software and follow along in the space provided.

Example 3 Translating Algebraic Expressions into English Phrases

Change each algebraic expression into an equivalent English phrase. In each case translate the variable as "a number."

a. $5x$

b. $2n + 8$

c. $3(a-2)$

Solution

Now You Try It!

Use the space provided to work out the solution to the next example.

Example A Identifying Properties of Addition and Multiplication

Change each algebraic expression into an equivalent English phrase.

a. $10x$

b. $4a + 7$

c. $7(n - 5)$

4.R.4 Exercises

Concept Check

True/False. Determine whether each statement is true or false. If a statement is false, explain how it can be changed so the statement will be true. (**Note:** There may be more than one acceptable change.)

1. The order in which the values are given is particularly important when working with subtraction and division problems.

2. "More than" and "increased by" are key phrases specifying the operation of subtraction.

3. Division is indicated by the phrase "five less than a number."

4. Key phrases for parentheses can be used to limit ambiguity in English phrases.

Practice

Write the algebraic expressions described by the English phrases. Choose your own variable.

5. six added to a number

6. twenty decreased by the product of four and a number

7. eighteen less than the quotient of a number and two

Translate each pair of English phrases into algebraic expressions. Notice the differences between the algebraic expressions and the corresponding English phrases.

8. a. six less than a number

   b. six less a number

9. a. six less than four times a number

   b. six less four times a number

Write the algebraic expression described by the English phrase using the given variables.

10. the cost of purchasing a fishing rod and reel if the rod costs x dollars and the reel costs $8 more than twice the cost of the rod

Translate each algebraic expression into an equivalent English phrase. (There may be more than one correct translation.)

11. $-9x$

12. $\dfrac{9}{x+3}$

Writing & Thinking

13. Explain why translating addition and multiplication problems from English into algebra may be easier than changing subtraction or division problems. (Consider the properties previously studied.)

14. Explain the difference between $5(n+3)$ and $5n+3$ when converting from algebra to English.

review concepts

CHAPTER 5.R

Polynomial Functions

5.R.1 Greatest Common Factor (GCF) of a Set of Terms

5.R.2 Factoring Trinomials by Grouping

5.R.3 Review of Factoring Techniques

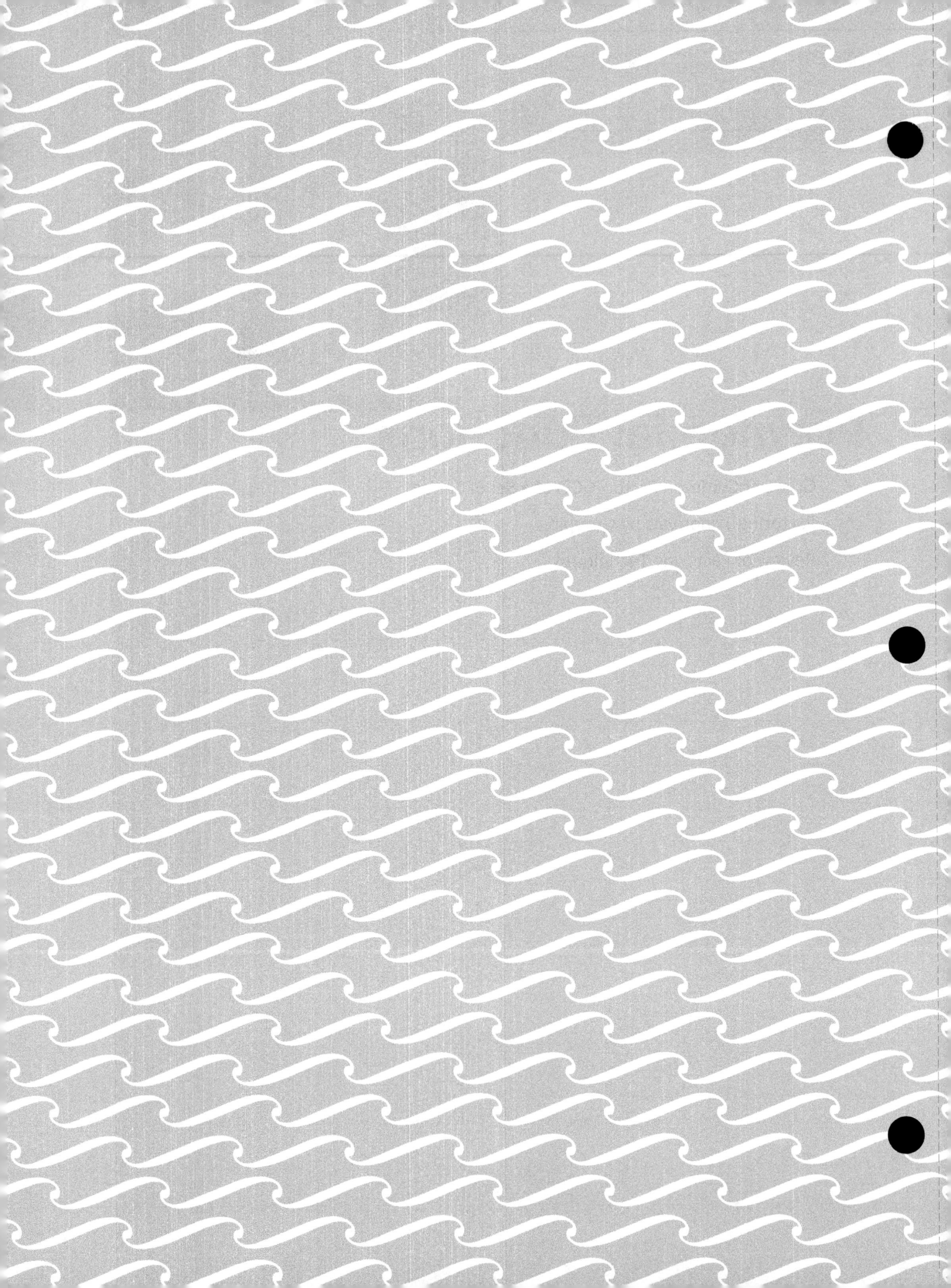

5.R.1 Greatest Common Factor (GCF) of a Set of Terms

1. The **greatest common factor** (**GCF**) of two or more integers is the _____

_____

Procedure for Finding the GCF of a Set of Terms

1. Find the prime factorization of _____

2. List all the factors that are _____

3. Raise each common factor to the _____

4. Multiply these powers to _____

Note: If there is no common prime factor or variable, then _____

PROCEDURE

▶ Watch and Work

Watch the video for Example 1 in the software and follow along in the space provided.

Example 1 Finding the GCF

Find the GCF for each set of algebraic terms.

a. $\{30, 45, 75\}$

b. $\{20x^4 y, 15x^3 y, 10x^5 y^2\}$

Solution

✏ Now You Try It!

Use the space provided to work out the solution to the next example.

Example A Factoring Polynomials by Grouping

Find the GCF for each set of algebraic terms.

a. $\{5,10,20\}$

b. $\{150xy, 250x^3y^2, 100x^2y^2\}$

5.R.1 Exercises

Concept Check

True/False. Determine whether each statement is true or false. If a statement is false, explain how it can be changed so the statement will be true. (**Note:** There may be more than one acceptable change.)

1. When finding the GCF of a polynomial, you need to consider only the coefficients.

2. One way to find the GCF of a set of numbers is to use the prime factorization of each number.

Practice

Find the GCF for each set of terms.

3. $\{25, 30, 75\}$

4. $\{8a^3, 16a^4, 20a^2\}$

5. $\{26ab^2, 39a^2b, 52a^2b^2\}$

6. $\{45x^2y^2z^2, 75xy^2z^3\}$

Applications

Solve.

7. *Halloween:* Bonnie volunteers to bring bags of candy to her child's class for the Halloween party this year. She buys one bag of candy A containing 150 pieces of candy, one bag of candy B containing 180 pieces of candy, and one bag of candy C containing 330 pieces of candy. She needs to use all the candy to create identical treat bags. How many treat bags can Bonnie make so that each one has the same number and variety of candy? How many of each type of candy will be in each bag?

Writing & Thinking

8. Explain why the GCF of $-3x^2 + 3$ is 3 and not -3.

5.R.2 Factoring Trinomials by Grouping

Analysis of Factoring by the ac-Method

General Method	Example
$ax^2 + bx + c$	$2x^2 + 9x + 10$

Step 1: Multiply _____

Multiply _____

Step 2: Find two integers whose product is ac and _____

Find two integers whose product is 20 and

_____

_____

_____

_____

Step 3: Rewrite the middle term (bx) using the _____

_____

_____

_____

_____

Step 4: Factor by grouping the _____

Factor by grouping the _____

_____

_____

Step 5: Factor out the common binomial factor. This will

Factor out the common binomial factor

give _____

_____

_____

PROCEDURE

▶ Watch and Work

Watch the video for Example 1 in the software and follow along in the space provided.

Example 1 Using the ac-Method

Use the ac-method to factor $3x^2 + 19x + 6$.

Solution

✏️ Now You Try It!

Use the space provided to work out the solution to the next example.

Example A Using the *ac*-Method

Use the *ac*-method to factor $3a^2 + 14a + 8$.

5.R.2 **Exercises**

Concept Check

True/False. Determine whether each statement is true or false. If a statement is false, explain how it can be changed so the statement will be true. (**Note:** There may be more than one acceptable change.)

1. A trinomial is factorable if the middle term is the difference of the inner and outer products of two binomials.

2. The first step in the ac-method of factoring is to rewrite the middle term.

3. Factoring can be checked by multiplying the factors and verifying that the product matches the original polynomial.

Practice

Completely factor each polynomial. If a polynomial cannot be factored, write "not factorable."

4. $6x^2 + 11x + 5$

5. $-x^2 + 3x - 2$

6. $x^2 + 8x + 64$

7. $9x^2 - 3x - 20$

8. $12x^2 - 38x + 20$

9. $5a^2 - 7a + 2$

Writing & Thinking

10. It is true that $2x^2 + 10x + 12 = (2x+6)(x+2) = (2x+4)(x+3)$. Explain how the trinomial can be factored in two ways. Is there some kind of error?

11. It is true that $5x^2 - 5x - 30 = (5x-15)(x+2)$. Explain why this is not the completely factored form of the trinomial.

5.R.3 Review of Factoring Techniques

General Guidelines for Factoring Polynomials

1. **Always look for a common** _____

   If the leading coefficient is negative, _____

2. **Check the number of terms.**

   a. **Two terms:**

      1. difference of _____

      2. sum of _____

      3. difference of _____

      4. sum of _____

   b. **Three terms:**

      1. _____

      2. _____

         Guidelines for the trial-and-error method

         a. If the sign of the constant term is positive, the signs in _____

            _____

         b. If the sign of the constant term is negative, the signs in _____

            _____

      3. _____

         Guidelines for the ac-method

         a. Multiply _____

         b. Find two integers whose product is ac and _____

            _____

         c. Rewrite the middle term (bx) using _____

         d. _____

   c. **Four terms:**

      Group terms with a common factor and _____

3. Check the possibility of _____

   Checking: Factoring can be checked by _____

   The product should be _____

5.R.3 Exercises

Concept Check

True/False. Determine whether each statement is true or false. If a statement is false, explain how it can be changed so the statement will be true. (**Note:** There may be more than one acceptable change.)

1. You should always start by checking the number of terms when factoring a polynomial.

2. If a trinomial is to be factored, the trial-and-error or *ac*-methods can be used.

3. If there are four terms in a polynomial, it cannot be factored.

Practice

Completely factor each of the given polynomials. If a polynomial cannot be factored, write "not factorable."

4. $x^2 - 100$

5. $x^2 + 10x + 25$

6. $x^2 + 16x + 64$

7. $20x^2 - 21x - 54$

8. $2y^2 + 6yz + 5y + 15z$

9. $x^3 + 125$

review concepts

CHAPTER 6.R

Rational Functions and Conic Sections

6.R.1 Introduction to Rational Expressions

6.R.2 Special Products of Binomials

6.R.3 Special Factoring Techniques

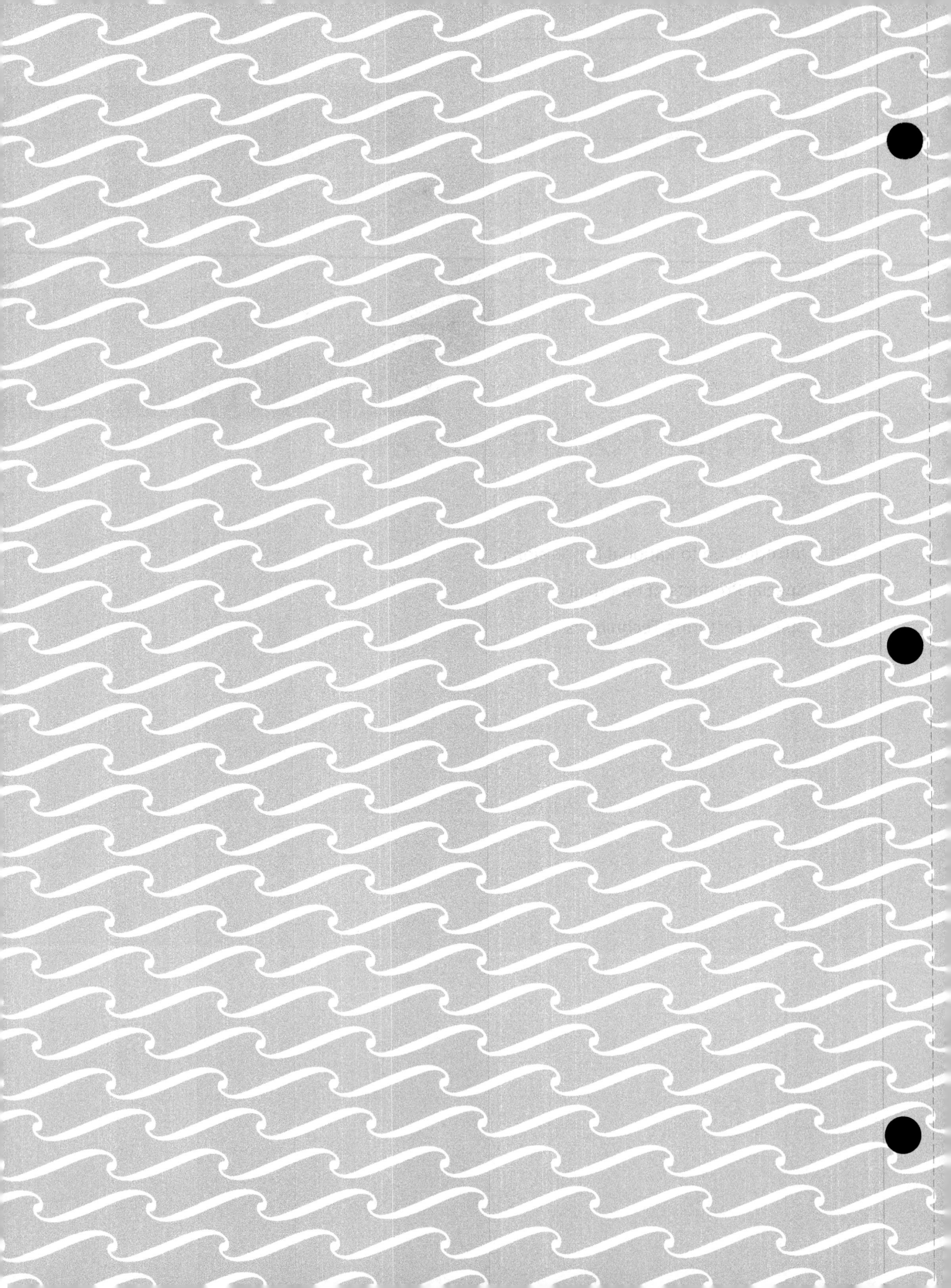

6.R.1 Introduction to Rational Expressions

Rational Expressions

A rational expression is an algebraic expression that can be written in the form

_____

DEFINITION

Summary of Arithmetic Rules for Rational Numbers (or Fractions)

A **fraction** (or **rational number**) is a number that can be written in the form

_____

The Fundamental Principle:

_____

The **reciprocal** of $\dfrac{a}{b}$ is _____

Multiplication:

_____

Division:

_____

Addition:

_____

Subtraction:

_____

PROPERTIES

The Fundamental Principle of Rational Expressions

If $\dfrac{P}{Q}$ is a rational expression and P, Q, and K are polynomials where

$$Q, K \neq 0, \text{ then}$$

_____

DEFINITION

▶ Watch and Work

Watch the video for Example 3 in the software and follow along in the space provided.

Example 3 Reducing Rational Expressions

Use the fundamental principle to reduce each expression to lowest terms. State any restrictions on the variable by using the fact that no denominator can be 0. This restriction applies to denominators **before and after** a rational expression is reduced.

a. $\dfrac{2x-10}{3x-15}$

b. $\dfrac{x^2-3x-4}{x^2-16}$

c. $\dfrac{y-10}{10-y}$

Solution

✏ Now You Try It!

Use the space provided to work out the solution to the next example.

Example A Reducing Rational Expressions

Reduce each expression to lowest terms. State any restrictions on the variable.

a. $\dfrac{2x-6}{5x-15}$

b. $\dfrac{x^2-x-20}{x^2-25}$

c. $\dfrac{x-5}{5-x}$

Opposites in Rational Expressions

For a polynomial P,

_____

In particular,

_____

DEFINITION

6.R.1 Exercises

Concept Check

True/False. Determine whether each statement is true or false. If a statement is false, explain how it can be changed so the statement will be true. (**Note:** There may be more than one acceptable change.)

1. A simplified rational expression cannot have any common factors other than 1 and -1 in both the numerator and denominator.

2. The difference between a rational number and a rational expression is that a rational expression generally has polynomials in the numerator and/or denominator.

3. While a rational number cannot have a zero denominator, a rational expression can have a zero denominator.

4. If a denominator is $x + 5$, it is defined for all values except 5.

Practice

Reduce each expression to lowest terms. State any restrictions on the variable(s).

5. $\dfrac{9x^2y^3}{12xy^4}$

7. $\dfrac{xy - 3y + 2x - 6}{y^2 - 4}$

6. $\dfrac{2x - 8}{16 - 4x}$

8. Evaluate $\dfrac{3y - 4}{y^2 + 25}$ for $y = 3$

Applications

Solve.

9. *Event Planning:* The cost of renting a party room with tables, chairs, and simple decorations is $200 plus $15 per person attending.

   a. Write a rational expression that represents the total price per person for renting the party room, where x is the number of people attending.

   b. What is the price per person to rent the party room if 10 people are attending?

   c. Determine which values of the variable will make the rational expression from Part **a.** undefined.

   d. Considering the context of the given problem, are there any additional restrictions on the variable? If so, explain why these restrictions are in place.

10. *Rectangles:* The area of a rectangle (in square feet) is represented by the polynomial function $A(x) = 4x^2 - 4x - 15$. If the length of the rectangle is $(2x+3)$ feet, find a representation for the width.

$$A(x) = 4x^2 - 4x - 15$$

$$2x + 3$$

Writing & Thinking

11. a. Define the term rational expression.

b. Give an example of a rational expression that is undefined for $x = -2$ and $x = 3$ and has a value of 0 for $x = 1$. Explain how you determined this expression.

c. Give an example of a rational expression that is undefined for $x = -5$ and never has a value of 0. Explain how you determined this expression.

6.R.2 Special Products of Binomials

Difference of Two Squares

_____

DEFINITION

Squares of Binomials (Perfect Square Trinomials)

$(x+a)^2 = x^2 + 2ax + a^2$ Square of a _____

$(x-a)^2 = x^2 - 2ax + a^2$ Square of a _____

DEFINITION

▶ Watch and Work

Watch the video for Example 3 in the software and follow along in the space provided.

Example 3 Squares of Binomials

Find the following products.

a. $(2x+3)^2$ b. $(5x-1)^2$ c. $(9-x)^2$ d. $(y^3+1)^2$

Solution

Now You Try It!

Use the space provided to work out the solution to the next example.

Example A Squares of Binomials

Find the following products.

a. $(3x+5)^2$

b. $(4x-2)^2$

c. $(8-2x)^2$

d. $(2y^3-2)^2$

6.R.2 **Exercises**

Concept Check

True/False. Determine whether each statement is true or false. If a statement is false, explain how it can be changed so the statement will be true. (**Note:** There may be more than one acceptable change.)

1. When two binomials are in the form of the sum and difference of the same term, the product will be a trinomial.

2. When the two binomials being multiplied together are the same, the product will be a trinomial.

3. Perfect square trinomials result from squaring a binomial sum or a binomial difference.

4. When finding the product of two binomials that are in the form of the sum and difference of the same two terms, the FOIL method and the difference of two squares formula will produce different results.

Practice

Find each product and identify any that are either the difference of two squares or a perfect square trinomial.

5. $(x-7)^2$

6. $(x+12)(x-12)$

7. $(3x-2)(3x-2)$

8. $(5x-9)(5x+9)$

Applications

Solve.

9. ***Geometry:*** A square is 20 inches on each side. A square x inches on each side is cut from each corner of the square.

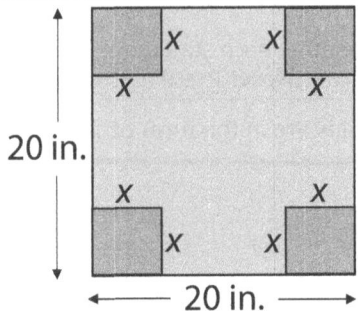

a. Represent the area of the remaining portion of the square in the form of a polynomial function $A(x)$.

b. Represent the perimeter of the remaining portion of the square in the form of a polynomial function $P(x)$.

10. ***Probability:*** In the case of binomial probabilities, if x is the probability of success in one trial of an event, then the expression $f(x) = 15x^4 (1-x)^2$ is the probability of 4 successes in 6 trials where $0 \le x \le 1$.

a. Represent the expression $f(x)$ as a single polynomial by multiplying the polynomials.

b. If a fair coin is tossed, the probability of heads occurring is $\frac{1}{2}$. That is, $x = \frac{1}{2}$. Find the probability of 4 heads occurring in 6 tosses.

Writing & Thinking

11. A square with sides of length $(x+5)$ can be broken up as shown in the diagram. The sums of the areas of the interior rectangles and squares is equal to the total area of the square: $(x+5)^2$. Show how this fits with the formula for the square of a sum.

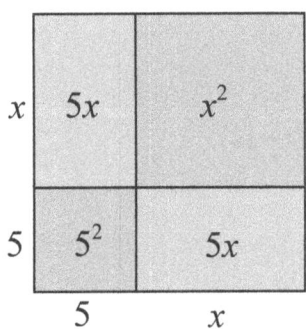

6.R.3 Special Factoring Techniques

Difference of Two Squares

Consider the polynomial $x^2 - 25$. By recognizing this expression as the **difference of two squares**, we can go directly to the factors:

$$x^2 - 25 = \underline{\hspace{3cm}}$$

DEFINITION

Sum of Two Squares

The sum of two squares is an expression of the form $\underline{\hspace{6cm}}$

DEFINITION

In a perfect square trinomial, both the _____ and _____ terms of the trinomial must be perfect squares. If the first term is of the form x^2 and the last term is of the form a^2, then the middle term must be of the form _____ or _____.

▶ Watch and Work

Watch the video for Example 3 in the software and follow along in the space provided.

Example 3 Factoring Perfect Square Trinomials

Factor completely.

a. $z^2 - 12z + 36$

b. $4y^2 + 12y + 9$

c. $2x^3 - 8x^2 y + 8xy^2$

d. $\left(x^2 + 6x + 9\right) - y^2$

Solution

✏ Now You Try It!

Use the space provided to work out the solution to the next example.

Example A Factoring Perfect Square Trinomials

Factor completely.

a. $z^2 + 40z + 400$

b. $y^2 - 14y + 49$

c. $3x^2z - 18xyz + 27y^2z$

d. $(y^2 + 8y + 16) - z^2$

6.R.3 **Exercises**

Concept Check

True/False. Determine whether each statement is true or false. If a statement is false, explain how it can be changed so the statement will be true. (**Note:** There may be more than one acceptable change.)

1. The expression $x^2 + 20x + 100$ is a perfect square trinomial.

2. When factoring polynomials, always look for a common monomial factor first.

3. The sum of two squares, $(x^2 + a^2)$, is factorable.

Practice

Completely factor each of the given polynomials. If a polynomial cannot be factored, write "not factorable."

4. $25 - z^2$

5. $y^2 - 16y + 64$

6. $x^2 + 64y^2$

Solve.

7. a. Represent the area of the shaded region of the square shown below as the difference of two squares.

    b. Use the factors of the expression in Part **a.** to draw (and label the sides of) a rectangle that has the same area as the shaded region.

Writing & Thinking

8. a. Show that the sum of the areas of the rectangles and squares in the figure is a perfect square trinomial.

 b. Rearrange the rectangles and squares in the form of a square and represent its area as the square of a binomial.

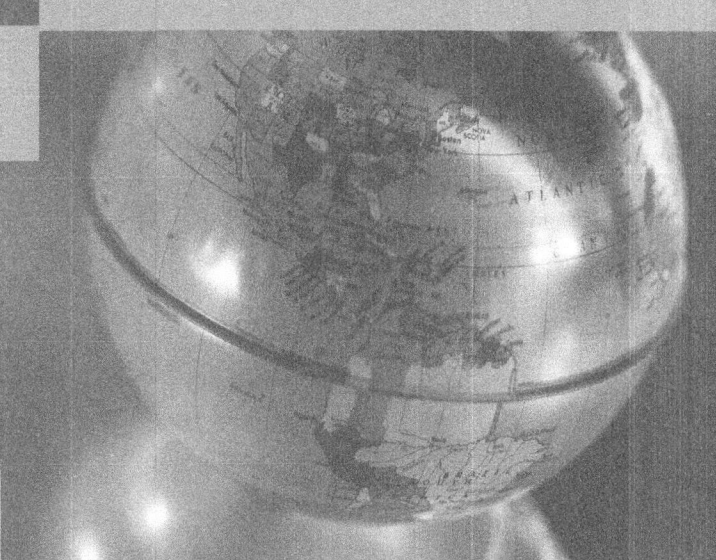

integrated review

CHAPTER 7.R

Exponential and Logarithmic Functions

7.R.1 Rules for Exponents

7.R.2 Power Rules for Exponents

7.R.3 Rational Exponents

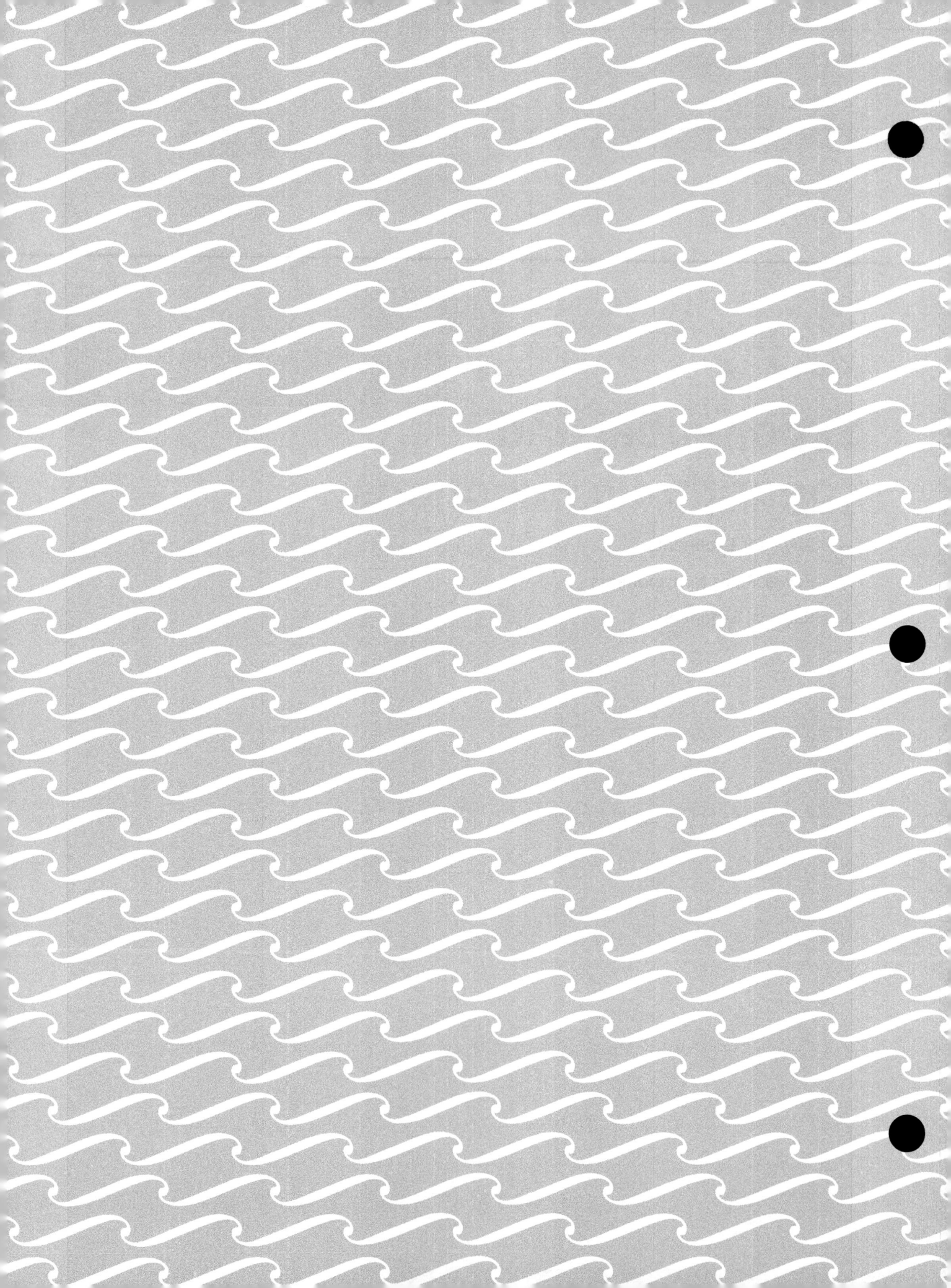

7.R.1 **Rules for Exponents**

The Product Rule for Exponents

If a is a nonzero real number and m and n are integers, then

_____

In words, to multiply powers with the same base, _____

PROPERTIES

The Exponent 0

If a is a nonzero real number, then _____

The expression _____ is _____

DEFINITION

Quotient Rule for Exponents

If a is a nonzero real number and m and n are integers, then

_____

In words, to divide two powers with the same base, _____

PROPERTIES

Rule for Negative Exponents

If a is a nonzero real number and n is an integer, then

_____

PROPERTIES

▶ Watch and Work

Watch the video for Example 6 in the software and follow along in the space provided.

Example 6 Negative Exponents

Use the rule for negative exponents to simplify each expression so that it contains only positive exponents.

  a. 5^{-1}

  b. x^{-3}

  c. $x^{-9} \cdot x^{7}$

Solution

✏ Now You Try It!

Use the space provided to work out the solution to the next example.

Example A Negative Exponents

Use the rule for negative exponents to simplify each expression so that it contains only positive exponents.

a. 7^{-1}

b. x^{-7}

c. $x^{-11} \cdot x^{6}$

Summary of the Rules for Exponents

For any nonzero real number a and integers m and n:

1. The exponent 1: _____

2. The exponent 0: _____

3. The product rule:_____

4. The quotient rule:
   _____

5. Negative exponents:
   _____

PROPERTIES

7.R.1 **Exercises**

Concept Check

True/False. Determine whether each statement is true or false. If a statement is false, explain how it can be changed so the statement will be true. (**Note:** There may be more than one acceptable change.)

1. If a constant does not have an exponent written, it is assumed that the exponent is 0.

2. If a is a nonzero real number and n is an integer, then $a^{-n} = -a^n$.

3. Since the product rule is stated for integer exponents, the rule is also valid for 0 and negative exponents.

4. When using the quotient rule, you should subtract the smaller exponent from the larger exponent.

Practice

Simplify each expression. The final form of the expressions with variables should contain only positive exponents. Assume that all variables represent nonzero numbers.

5. $y^3 \cdot y^8$

6. $\dfrac{y^7}{y^2}$

7. $x^{-3} \cdot x^0 \cdot x^2$

8. $\dfrac{10^4 \cdot 10^{-3}}{10^{-2}}$

Applications

Solve.

9. *Computers:* Rylee wants to move all her files to a new hard drive that has 2^{12} GB of storage on it. She wants to designate the same amount of storage for each of 2^4 projects. How much storage should be assigned to each project? Write your answer as a power of two.

10. *Bacteria:* Trey is studying patterns in bacteria. For a positive test result in his experiment, bacteria must grow in population at a minimum rate of 3^2 in 24 hours. If the initial population of the bacteria is 3^5 and his final measurement after 24 hours is 3^8, should he mark the test as positive or negative?

7.R.2 Power Rules for Exponents

Power Rule for Exponents

If a is a nonzero real number and m and n are integers, then

_____

In other words, the value of a power raised to a power can be found by _____

_____

PROPERTIES

Rule for Power of a Product

If a and b are nonzero real numbers and n is an integer then

_____

In words, a power of a product is found by _____

PROPERTIES

Rule for Power of a Quotient

If a and b are nonzero real numbers and n is an integer, then

_____

In words, a power of a quotient (in fraction form) is found by _____

_____

PROPERTIES

▶ Watch and Work

Watch the video for Example 5 in the software and follow along in the space provided.

Example 5 Using Two Approaches with Fractional Expressions and Negative Exponents

Simplify: $\left(\dfrac{x^3}{y^5}\right)^{-4}$

Solution

✏ Now You Try It!

Example A Using Two Approaches with Fractional Expressions and Negative Exponents

Simplify:

$$\left(\frac{x^6}{y^3}\right)^{-5}$$

Summary of the Rules for Exponents

For any nonzero real numbers a and b and integers m and n:

1. The exponent 1: _____

2. The exponent 0: _____

3. The product rule: _____

4. The quotient rule:
   _____

5. Negative exponents:
   _____

6. Power rule: _____

7. Power of a product: _____

8. Power of a quotient:
   _____

PROPERTIES

7.R.2 **Exercises**

Concept Check

True/False. Determine whether each statement is true or false. If a statement is false, explain how it can be changed so the statement will be true. (**Note:** There may be more than one acceptable change.)

1. Taking the reciprocal of a fraction changes the sign of any exponent in the fraction.

2. For an exponent to refer to -7 as the base, -7 must be in parentheses.

3. When simplifying an expression with exponents, the rules for exponents must be used in a specific order or the answer will vary.

4. The expression -8^2 simplifies to -64.

Practice

Use the rules for exponents to simplify each of the expressions. Assume that all variables represent nonzero real numbers.

5. $\left(2^{-3}\right)^{-2}$

6. $-3\left(7xy^2\right)^0$

7. $-2\left(3x^5 y^{-2}\right)^{-3}$

8. $\left(\dfrac{x}{2}\right)^3$

9. $\left(\dfrac{2x^2 y}{y^3}\right)^{-4}$

10. $\left(\dfrac{5a^4 b^{-2}}{6a^{-4} b^3}\right)^{-2}\left(\dfrac{5a^3 b^4}{2^{-2} a^{-2} b^{-2}}\right)^3$

7.R.3 Rational Exponents

Radical Notation

If n is an integer greater than 1, then _____

The expression $\sqrt[n]{a}$ is called _____

The symbol $\sqrt[n]{}$ is called _____

n is called _____

_____

Note: If no index is given, it is _____

For example, _____

DEFINITION

The General Form $a^{\frac{m}{n}}$

If n is an integer greater than 1, m is any integer, and $a^{\frac{1}{n}}$ is a real number, then

_____

In radical notation: _____

DEFINITION

▶ Watch and Work

Watch the video for Example 4 in the software and follow along in the space provided.

Example 4 Simplifying Radical Notation by Changing to Exponential Notation

Simplify each expression by first changing it into an equivalent expression with rational exponents. Then rewrite the answer in simplified radical form.

a. $\sqrt[4]{\sqrt[3]{x}} =$

b. $\sqrt[3]{a}\sqrt{a} =$

c. $\dfrac{\sqrt{x^3}\,\sqrt[3]{x^2}}{\sqrt[5]{x^2}} =$

✏ Now You Try It!

Use the space provided to work out the solution to the next example.

Example A Simplifying Radical Notation by Changing to Exponential Notation

Rewrite the expression in simplified radical form.

a. $\sqrt[5]{\sqrt[2]{x}}$

b. $\sqrt[4]{x} \cdot \sqrt{x}$

c. $\dfrac{\sqrt[3]{x^2} \cdot \sqrt{x^3}}{\sqrt[6]{x^5}}$

7.R.3 **Exercises**

Concept Check

True/False. Determine whether each statement is true or false. If a statement is false, explain how it can be changed so the statement will be true. (**Note:** There may be more than one acceptable change.)

1. The same rules for exponents apply to both integer exponents and rational exponents.

2. If the cube root of 7 were to be converted into exponential notation it would be $\sqrt[3]{7}$.

3. Any expression to the power 0, such as $\left(\sqrt[4]{x}\right)^0$, is equal to 1.

4. The expression $y^{\frac{1}{2}}$ can be rewritten in radical notation as $\sqrt{y^2}$.

Practice

5. Use radical notation to write an expression that is equivalent to $8^{\frac{1}{3}}$.

6. Use exponential notation to write an expression that is equivalent to $\sqrt{3}$.

Simplify each numerical expression.

7. $100^{-\frac{1}{2}}$

8. $64^{\frac{2}{3}}$

9. Simplify $\dfrac{a^{\frac{1}{2}} \cdot a^{-\frac{3}{4}}}{a^{-\frac{1}{2}}}$. Assume that all variables represent positive real.

10. Simplify $\dfrac{\sqrt[4]{y^3}}{\sqrt[6]{y}}$ by first changing it into an equivalent expression with rational exponents. Rewrite the

answer in simplified radical form. Assume that all variables represent positive real numbers.

Applications

Solve.

11. *Area:* The width of a rectangle is $\sqrt[3]{64^2}$ ft and the length is $216^{\frac{2}{3}}$ ft. What is the area of the rectangle?

12. *Amusement Parks:* An amusement park is creating signs to indicate the velocity of the roller coaster car on certain hills of the most popular rides. A roller coaster car gains kinetic energy as it goes down a hill.

The velocity, or speed, of an object in kilometers per hour (kph) can be determined by $V = \left(\dfrac{2k}{m}\right)^{\frac{1}{2}}$, where k

is the kinetic energy of the object in joules (J) and m is the mass of the object in kilograms (kg).

  a. For the most popular roller coaster, the car has a mass of 300 kg and the car has a kinetic energy of 375,000 J on the first hill. What velocity does the car obtain on the first hill?

  b. For the second most popular roller coaster, the car has a mass of 350 kg and the car has a kinetic energy of 70,000 on the first hill. What velocity does the car obtain on the first hill?

Writing & Thinking

13. Is $\sqrt[5]{a} \cdot \sqrt{a}$ the same as $\sqrt[5]{a^2}$? Explain why or why not.

14. Assume that x represents a positive real number. Describe what kind of number the exponent n must be for x^n to mean

  a. a product.

  b. a quotient.

  c. 1.

  d. a radical state.

integrated review

CHAPTER 8.R

Systems of Equations

8.R.1 Systems of Linear Equations: Solutions by Graphing

8.R.2 Systems of Linear Inequalities

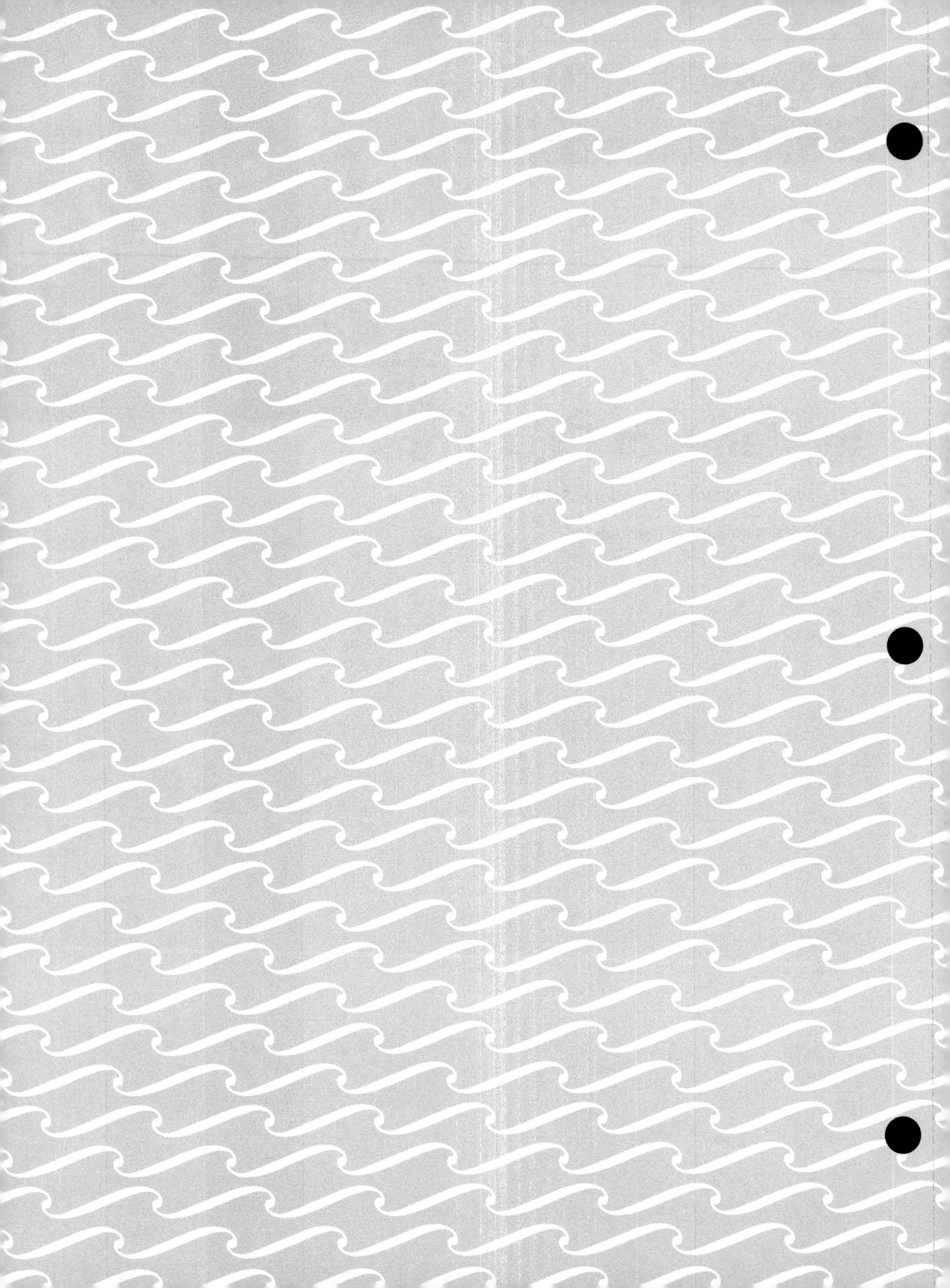

8.R.1 Systems of Linear Equations: Solutions by Graphing

1. Form a system of linear equations or _____

2. The **solution of a system** of linear equations is the set of ordered pairs (or points) that satisfy _____

_____

To Solve a System of Linear Equations by Graphing

1. Graph both linear equations on _____

2. Observe the point of _____
   a. If the slopes of the two lines are different, then _____
      The system has _____
   b. If the lines have the same slope and different y-intercepts, then _____
      The system has _____
   c. If the lines are the same line, then _____
      _____

3. Check the solution (if there is one) in _____

PROCEDURE

Consistent and Inconsistent Systems of Linear Equations

1. A system is **consistent** if _____

2. A system is **inconsistent** _____

DEFINITION

▶ Watch and Work

Watch the video for Example 3 in the software and follow along in the space provided.

Example 3 Solving Systems (One Solution/A Consistent System)

Solve the system of equations by graphing.

$$\begin{cases} y = -x + 4 \\ y = 2x + 1 \end{cases}$$

Solution

✏ Now You Try It!

Use the space provided to work out the solution to the next example.

Example A Solving Systems (One Solution/A Consistent System)

Solve the system of equations by graphing.

$$\begin{cases} y = 2x + 6 \\ y = -x - 3 \end{cases}$$

Dependent and Independent Systems of Linear Equations

If the graphs of two linear equations are

a. the same line, then the _____

b. different lines, then the _____

DEFINITION

8.R.1 Exercises

Concept Check

True/False. Determine whether each statement is true or false. If a statement is false, explain how it can be changed so the statement will be true. (**Note:** There may be more than one acceptable change.)

1. To check a solution, substitute it into one of the equations. If the solution satisfies one equation it will satisfy all of the equations.

2. A system of equations with graphs that are parallel lines has exactly one solution.

3. A system of equations with graphs that intersect at one point has exactly one solution.

4. A system of equations with graphs that are the same line has infinitely many solutions.

Practice

Determine which of the given points, if any, lie on both of the lines in the systems of equations by substituting each point into both equations.

5. $\begin{cases} 2x + 4y - 6 = 0 \\ 3x + 6y - 9 = 0 \end{cases}$

   a. $(1, 1)$

   b. $(2, 0)$

   c. $\left(0, \dfrac{3}{2}\right)$

   d. $(-1, 3)$

The graphs of the lines represented by system of equations are given. Determine the solution of the system by looking at the graph. Check your solution by substituting into both equations.

6. $\begin{cases} x + 2y = 4 \\ x - y = -2 \end{cases}$

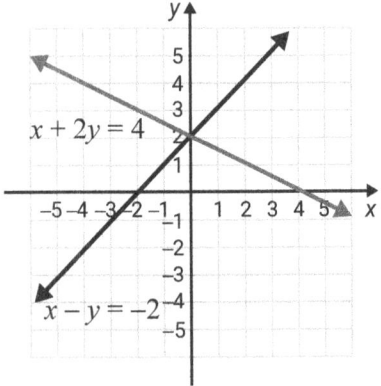

Solve each system of equations by graphing.

7. $\begin{cases} x - 2y = 4 \\ x = 4 \end{cases}$

8. $\begin{cases} 2x + y = 0 \\ 4x + 2y = -8 \end{cases}$

Applications

Each of the following applications has been modeled using a system of equations. Solve the system graphically.

9. *Swimming Pools:* OSHA recommends that swimming pool owners clean their pool decks with a solvent composed of a 12% chlorine solution and a 3% chlorine solution. Fifteen gallons of the solvent consists of 6% chlorine. How much of each of the mixing solutions were used?

Let x = the number of gallons of the 12% solution
and y = the number of gallons of the 3% solution.

The corresponding modeling system is $\begin{cases} x + y = 15 \\ 0.12x + 0.03y = 0.06(15) \end{cases}$

10. *School Supplies:* A student bought a calculator and a textbook for a course in algebra. He told his friend that the total cost was $170 (without tax) and that the calculator cost $20 more than twice the cost of the textbook. What was the cost of each item?

Let x = the cost of the calculator
and y = the cost of the textbook.

The corresponding modeling system is $\begin{cases} x + y = 170 \\ x = 2y + 20 \end{cases}$

Writing & Thinking

11. Explain, in your own words, why the answer to a consistent system of linear equations can be written as an ordered pair.

8.R.2 Systems of Linear Inequalities

To Solve a System of Two Linear Inequalities

1. For each inequality, graph the boundary line and _____

2. Determine the region of the graph that is _____
   _____

   This region is called the _____
   _____

3. To check, pick one test-point in the _____

Note: If there is no intersection, then the system has no solution.

PROCEDURE

▶ Watch and Work

Watch the video for Example 1 in the software and follow along in the space provided.

Example 1 Solving Systems of Linear Inequalities

Solve the system of linear inequalities graphically. $\begin{cases} x \leq 2 \\ y \geq -x+1 \end{cases}$

Solution

✏ Now You Try It!

Use the space provided to work out the solution to the next example.

Example A Solving Systems of Linear Inequalities

Solve the system of linear inequalities graphically.

$$\begin{cases} y \geq 2 \\ x - y < 4 \end{cases}$$

Possible Solutions to Systems of Linear Inequalities

When the boundary lines are parallel there are three possibilities:

1. The common region will be in the form of _____

2. The common region will be a _____
   _____

3. There will be _____

8.R.2 **Exercises**

Concept Check

True/False. Determine whether each statement is true or false. If a statement is false, explain how it can be changed so the statement will be true. (**Note:** There may be more than one acceptable change.)

1. When boundary lines are parallel, the system of linear inequalities has no solution.

2. If two half-planes overlap, that region is the union of the graphs.

3. Half-planes are the graphs of linear inequalities.

4. If the graphs of two linear inequalities have no intersection, then the system has no solution.

Practice

Solve the systems of two linear inequalities graphically.

5. $\begin{cases} y > 2 \\ x \geq -3 \end{cases}$

7. $\begin{cases} 2x - 3y \geq 0 \\ 8x - 3y < 36 \end{cases}$

6. $\begin{cases} y > 3x + 1 \\ -3x + y < -1 \end{cases}$

8. $\begin{cases} y > x - 4 \\ y < x + 2 \end{cases}$

Applications

Solve.

9. *Fundraising:* Robin is planning a charity ball to raise money for her favorite charity. There are two different ticket options. The VIP option includes dinner, dancing, and cocktails for $150 per ticket. The regular option includes dancing and cocktails for $75 per ticket. Robin wants to make at least $14,000 in ticket sales. The ballroom that is being used for the charity event has a maximum capacity of 150 people.

   a. Write two linear inequalities to describe the situation. Let the variable x represent the number of VIP tickets sold and let the variable y represent the number of regular tickets sold.

   b. Graph the two linear inequalities on the same coordinate plane.

   c. Describe the solution set for the situation.

   d. Can Robin reach her sales goal if she only sells tickets for the regular option? Explain why or why not.

Writing & Thinking

10. Graph the inequalities and explain how you can tell that there is no solution.
$$\begin{cases} y \le 2x - 5 \\ y \ge 2x + 3 \end{cases}$$

Math@Work

Math@Work

Basic Inventory Management

As an business manager you will need to evaluate the company's inventory several times per year. While evaluating the inventory, you will need to ensure that enough of each product will be in stock for future sales based on current inventory count, predicted sales, and product cost. Let's say that you check the inventory four times a year, or quarterly. You will be working with several people to get all of the information you need to make the proper decisions. You need the sales team to give you accurate predictions of how much product they expect to sell. You need the warehouse manager to keep an accurate count of how much of each product is currently in stock and how much of that stock has already been sold. You will also have to work with the product manufacturer to determine the cost to produce and ship the product to your company's warehouse. It's your job to look at this information, compare it, and decide what steps to take to make sure you have enough of each product in stock for sales needs. A wrong decision can potentially cost your company a lot of money.

Suppose you get the following reports: an inventory report of unsold products from the warehouse manager and the report on predicted sales for the next quarter (three months) from the sales team.

Unsold Products	
Item	Number in Stock
A	5025
B	150
C	975
D	2150

Predicted Sales	
Item	Expected Sales
A	4500
B	1625
C	1775
D	2000

Suppose the manufacturer gives you the following cost list for the production and shipment of different amounts of each inventory item.

Item	Amount	Cost	Amount	Cost	Amount	Cost
A	500	$875	1000	$1500	1500	$1875
B	500	$1500	1000	$2500	1500	$3375
C	500	$250	1000	$400	1500	$525
D	500	$2500	1000	$4250	1500	$5575

1. Which items and how much of each item do you need to purchase to make sure the inventory will cover the predicted sales?

2. If you purchase the amounts from Problem 1, how much will this cost the company?

3. By ordering the quantities you just calculated, you are ordering the minimum of each item to cover the expected sales. If the actual sales during the quarter are higher than expected, what might happen? How would you handle this situation?

4. Which skills covered in this chapter were necessary to help you make your decisions?

Math@Work

Hospitality Management: Preparing for a Dinner Service

As the manager of a restaurant you will need to make sure everything is in place for each meal service. This means that you need to predict and prepare for busy times, such as a Friday night dinner rush. To do this, you will need to obtain and analyze information to determine how much of each meal is typically ordered. After you estimate the number of meals that will be sold, you need to communicate to the chefs how much of each item they need to expect to prepare. An additional aspect of the job is to work with the kitchen staff to make sure you have enough ingredients in stock to last throughout the meal service.

You are given the following data, which is the sales records for the signature dishes during the previous four Friday night dinner services.

Week	Meal A	Meal B	Meal C	Meal D
1	30	42	28	20
2	35	38	30	26
3	32	34	26	26
4	30	32	28	22

Meal C is served with a risotto, a type of creamy rice. The chefs use the following recipe, which makes 6 servings of risotto, when they prepare Meal C. (**Note:** The abbreviation for tablespoon is T and the abbreviation for cup is c.)

$5\frac{1}{2}$ c chicken stock $2\frac{1}{3}$ T chopped shallots $\frac{1}{2}$ c red wine

$1\frac{1}{2}$ c rice   2 T chopped parsley $4\frac{3}{4}$ c thinly sliced mushrooms

2 T butter   2 T olive oil $\frac{1}{2}$ c Parmesan cheese

1. For the past four Friday night dinner services, what was the average number of each signature meal served? If the average isn't a whole number, explain why you would round this number either up or down.

2. Based on the average you obtained for Meal C, calculate how much of each ingredient your chefs will need to make the predicted amount of risotto.

3. The head chef reports the following partial inventory: $10\frac{3}{4}$ c rice, $15\frac{3}{4}$ c mushrooms, and 10 T shallots. Do you have enough of these three items in stock to prepare the predicted number of servings of risotto?

4. Which skills covered in this and the previous chapter helped you make your decisions?

Math@Work

Bookkeeper

As a bookkeeper, you will often receive bills and receipts for various purchases or expenses from employees of the company you work for. You will need to split the bill by expense code, assign costs according to customer, and reimburse an employee for their out-of-pocket spending. To do this you will need to know the company's reimbursement policies, the expense codes for different spending categories, and which costs fall into a particular expense category.

Suppose two employees from the sales department recently completed sales trips. Employee 1 flew out of state and visited two customers, Customer A and Customer B. This employee had a preapproved business meal with Customer B and was traveling for three days. Employee 2 drove out of state to visit Customer C. This employee stayed at a hotel for the night and then drove back the next day. The expenses for the two employees are as follows.

Employee 1	
Flight and Rental Car	$470.50
Hotel	$278.88
Meals	$110.56
Business Meal	$102.73
Presentation Materials	$54.86

Employee 2	
Miles Driven	578.5 miles
Fuel	$61.35
Hotel	$79.60
Meals	$53.23
Presentation Materials	$67.84

The expense categories used by your company to track spending are: Travel (includes hotel, flights, mileage, etc.), Meals (business), Meals (travel), and Supplies. Traveling employees are reimbursed up to $35 per day for meals while traveling and for all preapproved business meals. They also receive $0.565 per mile driven with their own car.

1. How much will you reimburse each employee for travel meals? Did either employee go over their allowed meal reimbursement amount?

2. What were the total expenses for each employee?

3. The company you work for keeps track of how much is spent on each customer. When a sales person visits multiple customers during one trip, the tracked costs are split between the customers. Fill in this table according to how much was spent on each customer for the different expense categories. (**Note:** For meals, only include the amount the employee was reimbursed.)

Expense	Customer A	Customer B	Customer C
Travel			
Meals (business)			
Meals (travel)			
Supplies			
Total			

Math@Work

Pediatric Nurse

As a pediatric nurse working in a hospital setting, you will be responsible for taking care of several patients during your work day. You will need to administer medications, set IVs, and check each patient's vital signs (such as temperature and blood pressure). While doctors prescribe the medications that nurses need to administer, it is important for nurses to double check the dosage amounts. Administering the incorrect amount of medication can be detrimental to the patient's health.

During your morning nursing round, you check in on three new male patients and obtain the following information.

	Patient A	Patient B	Patient C
Age	10	9	12
Weight (pounds)	81	68.5	112
Blood Pressure	97/58	100/59	116/73
Temperature (°F)	99.7	97.3	101.4
Medication	A	B	A

The following table shows the bottom of the range for abnormal blood pressure (BP) for boys. If either the numerator or the denominator of the blood pressure ratio is greater than or equal to the values in the chart, this can indicate a stage of hypertension.

Abnormal Blood Pressure for Boys by Age	
	Systolic BP / Diastolic BP
Age 9	109/72
Age 10	111/73
Age 11	113/74
Age 12	115/74

Source: http://www.nhlbi.nih.gov/health/public/heart/hbp/bp_child_pocket/bp_child_pocket.pdf

Medication Directions	
Medication	Dosage Rate
A	40 mg per 10 pounds
B	55 mg per 10 pounds

1. Do any of the patients have a blood pressure which may indicate they have hypertension? If yes, which patient(s)?

2. Use proportions to determine the amount of medication that should be administered to each patient based on weight. Round to the nearest 10 pounds before calculating.

3. The average body temperature is 98.6 degrees Fahrenheit. You are supposed to alert the doctor on duty if any of the patients have a temperature 2.5 degrees higher than average. For which patients would you alert a doctor?

4. Which skills covered in this chapter and the previous chapters were necessary to help you make your decisions?

Math@Work

Architecture

As a project architect, you will be part of a team that creates detailed drawings of the project that will be used during the construction phase. It will be your job to ensure that the project will meet guidelines given to you by your company, such as square footage requirements and budget constraints. You will also need to meet the design requirements requested by the client.

Suppose you are part of a team that is designing an apartment building. You are given the task to create the floor plan for an apartment unit with two bedrooms and one bathroom. The apartment management company that has contracted your company to do the project has several requirements for this specific apartment unit.

1. One bedroom is the "master bedroom" and must have at least 60 square feet more than the other bedroom.

2. All walls must intersect or touch at 90 degree angles.

3. The kitchen must have an area of no more than 110 square feet.

4. The apartment must be between 1000 square feet and 1050 square feet.

A preliminary sketch of the apartment is shown here.

1. Does the apartment have the required total square footage that was requested? Is it over or under the total required?

2. Does the apartment blueprint meet the other requirements given by the client? If not, what does not meet the requirements?

3. For this specific apartment unit, the total construction cost per square foot is estimated to be $75.75. Approximately how much will it cost to construct each two-bedroom apartment based on the floor plan?

Math@Work

Statistician: Quality Control

Suppose you are a statistician working in the quality control department of a company that manufactures the hardware sold in kits to assemble book shelves, TV stands, and other ready-to-assemble furniture pieces. There are three machines that produce a particular screw and each machine is sampled every hour. A measurement of the screw length is determined with a micrometer, which is a device used to make highly precise measurements. The screw is supposed to be 3 inches in length and can vary from this measurement by no more than 0.1 inches or it will not fit properly into the furniture. The following table shows the screw length measurements (in inches) taken each hour from each machine throughout the day. The screw length data from each machine has also been plotted

Screw Length Measurements (in inches)			
Sample Time	Machine A	Machine B	Machine C
8 a.m.	2.98	2.92	2.99
9 a.m.	3.00	2.94	3.00
10 a.m.	3.02	2.97	3.01
11 a.m.	2.99	2.96	3.03
12 p.m.	3.01	2.94	3.05
1 p.m.	3.00	2.95	3.04
2 p.m.	2.97	2.93	3.06
3 p.m.	2.99	2.92	3.08
Mean			
Range			

1. Calculate the mean and range of the data for each machine and place them in the bottom two rows of the table.

2. If the screw length can vary from 3 inches by no more than 0.1 inch (plus or minus), what are the lowest and highest values for length that will be acceptable? Place a horizontal line on the graph at each of these values on the vertical axis. These are the tolerance or specification limits for screw length.

3. Have any of the three machines produced an unacceptable part today? Are any of the machines close to making a bad part? If so, which one(s)?

4. Look at the graph and the means from the table that show the average screw length produced by each machine. Draw a bold horizontal line on the graph at 3 to emphasize the target length. Do all the machines appear to be making parts that vary randomly around the target of 3 inches?

5. Look at the range values from the table. Do any of the machines appear to have more variability in the length measurements than the others?

6. In your opinion, which machine is performing best? Would you recommend that any adjustments be made to any of the machines? If so, which one(s) and why?

Math@Work

Dental Assistant

As a dental assistant, your job duties will vary depending on where you work. Suppose you work in a dental office where you assist with dental procedures and managing patients' accounts. When a patient arrives for their appointment, you will need to review their chart and make sure they are up to date on preventive care, such as X-rays and cleanings. When the patient leaves, you will need to fill out an invoice to determine how much to charge the patient for their visit.

Dental patients generally have a new X-ray taken yearly. Cleanings are performed every 6 months, although some patients have their teeth cleaned more often. The following table shows the date of the last X-ray and cleaning for three patients that are visiting the office today. (**Note:** All dates are within the past year.)

Patient Histories		
Patient	**Last X–ray**	**Last Cleaning**
A	April 15	October 20
B	June 6	January 12
C	October 27	October 27

During Patient A's visit, she received a fluoride treatment and a cleaning. Patient A has no dental insurance. During Patient B's visit, he received a filling on one surface of a tooth. Patient B has dental insurance which pays for 60% of the cost of fillings. During Patient C's visit, he had a cleaning, a filling on one surface of a tooth, and a filling on two surfaces of another tooth. Patient C has dental insurance which covers the full cost of cleanings and 50% of the cost of fillings.

Fee Schedule	
Procedure	**Cost**
Cleaning	$95
Fluoride treatment	$35
Filling, One surface	$175
Filling, Two surfaces	$235
X–ray, Panoramic	$110

1. Using today's date, determine which of the three patients are due for a dental cleaning in the next two months?

2. Using today's date, determine which of the patients will require a new set of X-rays during this visit.

3. Determine the amount each patient will be charged for their visit (without insurance). Don't forget to include the cost of any X-rays that are due during the visit.

4. Use the insurance information to determine the amount that each patient will pay out-of-pocket at the end of their visit.

Math@Work

Financial Advisor

As a financial advisor working with a new client, you must first determine how much money your client has to invest. The client may have a lump sum that they have saved or inherited, or they may wish to contribute an amount monthly from their current salary. In the latter case, you must then have the client do a detailed budget, so that you can determine a reasonable amount that the client can afford to set aside on a monthly basis for investment.

The second piece of information necessary when dealing with a new client is determining how much risk-tolerance they have. If the client is young or has a lot of money to invest, they may be willing to take more risk and invest in more aggressive, higher interest-earning funds. If the client is older and close to retirement, or has little money to invest, they may prefer less-aggressive investments where they are essentially guaranteed a certain rate of return. The range of possible investments that would suit each client's needs and goals are determined using a survey of risk-tolerance.

Suppose you have a client who has a total of $25,000 to invest. You determine that there are two investment funds that meet the client's investment preferences. One option is an aggressive fund that earns an average of 12% interest and the other is a more moderate fund that earns an average of 5% interest. The client desires to earn $2300 this year from these investments.

Investment Type	Principal Invested	·	Interest Rate	=	Interest Earned
Aggressive Fund	x				
Moderate Fund					

To determine the amount of interest earned you know to use the table above and the formula $I = Prt$, where I is the interest earned, P is the principal or amount invested, r is the average rate of return, and t is the length of time invested. Since the initial investment will last one year, $t = 1$.

1. Fill in the table with the known information. If x is the amount invested in the aggressive fund and the total amount to be invested is $25,000, create an expression involving x for the amount that will be left to invest in the moderate fund. Place this expression in the appropriate cell of the table.

2. Determine an expression in x for the interest earned on each investment type by multiplying the principal by the interest rate.

3. Determine the amount invested in each fund by setting up an equation using the expressions in column four and the fact that the client desires to earn $2300 from the interest earned on both investments.

4. Verify that the investment amounts calculated for each fund in the previous step are correct by calculating the actual interest earned in a year for each and making sure they sum to $2300.

5. Why would you not advise your client to invest all their money in the fund earning 12% interest, after all, it has the highest average interest rate?

Math@Work
Market Research Analyst

As a market research analyst, you may work alone at a computer, collecting and analyzing data, and preparing reports. You may also work as part of a team or work directly with the public to collect information and data. Either way, a market research analyst must have strong math and analytical skills and be very detail-oriented. They must have strong critical-thinking skills to assess large amounts of information and be able to develop a marketing strategy for the company. They must also possess good communication skills in order to interpret their research findings and be able to present their results to clients.

Suppose you work for a shoe manufacturer who wants to produce a new type of lightweight basketball sneaker similar to a product a competitor recently released into the market. You have gathered some sales data on the competitor in order to determine if this venture would be worthwhile, which is shown in the table below. To begin your analysis, you create a scatter plot of the data to see the sales trend. (A scatter plot is a graph made by plotting ordered pairs in a coordinate plane in order to show the relationship between two variables.) You determine that the x-axis will represent the number of weeks after the competitors new sneaker went on the market and the y-axis will represent the amount of sales in thousands of dollars.

Number of Weeks x	Sales (in 1000s) y
3	15
6	22
9	28
12	35
15	43

1. Create a scatter plot of the sales data by plotting the ordered pairs in the table on the coordinate plane. Does the data on the graph appear to follow a linear pattern? If so, sketch a line that you feel would "best" fit this set of data. (A market research analyst would typically use computer software to perform a technique called regression analysis to fit a "best" line to this data.)

2. Using the ordered pairs corresponding to weeks 9 and 15, find the equation of a line running through these two data points.

3. Interpret the value calculated for the slope of the equation in Problem 2 as a rate of change in the context of the problem. Write a complete sentence.

4. If you assume that the sales trend in sneaker sales follows the model determined by the linear equation in Problem 2, predict the sneaker sales in 6 months. Use the approximation that 1 month is equal to 4 weeks.

5. Give at least two reasons why the assumption made in Problem 4 may be invalid?

Math@Work

Chemistry

As a pharmaceutical chemist, you will need an advanced degree in pharmaceutical chemistry, which combines biology, biochemistry, and pharmaceuticals. In this career, you will most likely spend your day in a lab setting creating new medications or researching their effectiveness. You will often work as part of a team working towards a joint goal. As a result, in addition to strong math skills and an understanding of chemistry, you will need to have good communication and leadership skills. Since you will be working directly with chemicals, you will also need to have a strong understanding of lab safety rules to ensure the safety of not only yourself, but your coworkers as well.

Suppose you work at a pharmaceutical company which creates and produces medications for various skin conditions. You are currently on a team which is developing an acne-controlling facial cleanser. Your team is working on determining the gentlest formula possible that is still effective so that the cleanser can be used on sensitive skin. Half of your team is working with salicylic acid and the other half is working with benzoyl peroxide.

As a part of your work, you will need to keep up on current research. Learning about new chemicals, new methods, and new research will be a continuous part of your life.

1. Perform an Internet search for benzoyl peroxide. How does it work to clean skin and prevent acne?

2. Perform an Internet search for salicylic acid. How does it work to clean skin and prevent acne?

3. Based on your research, which chemical seems better suited to treat acne on sensitive skin?

Another aspect of your career will involve the mixing of chemicals to create new compounds. Having the correct concentrations of chemicals is also important so the resulting solution works as you expect it to. When you don't have the correct concentration of a chemical in stock, it is possible to mix two concentrations together to obtain the desired concentration.

4. Your team wants to create a cleanser with 4% benzoyl peroxide. The lab currently has 2.5% and 10% concentrations of benzoyl peroxide in stock. To create 500 mL of 4% benzoyl peroxide, how much of each concentration should be combined?

Math@Work

Astronomy

Astronomy is the study of celestial bodies, such as planets, asteroids, and stars. While you work in the field of astronomy, you will use knowledge and skills from several other fields, such as mathematics, physics, and chemistry. An important tool of astronomers is the telescope. Several powerful telescopes are housed in observatories around the world. One of the many things astronomers use observatories for is discovering new celestial objects such as a near-Earth object (NEO). NEOs are comets, asteroids, and meteoroids that orbit the sun and cross the orbital path of Earth. The danger presented by NEOs is that they may strike the Earth and result in global catastrophic damage. (**Note:** The National Aeronautics and Space Administration (NASA) keeps track of all NEOs which are a potential threat at the website http://neo.jpl.nasa.gov/risk/)

For an asteroid to be classified as an NEO, the asteroid must have an orbit that partially lies within 0.983 and 1.3 astronomical units (AU) from the sun, where 1 AU is the furthest distance from the Earth to the sun, approximately 9.3×10^7 miles.

Near-Earth Object Distance			
	Minimum		Maximum
Distance in AU	0.983 AU	1 AU	1.3 AU
Distance in Miles		9.3×10^7 miles	

Suppose you discover three asteroids that you suspect may be NEOs. You perform some calculations and come up with the following facts. The furthest that Asteroid A is ever from the sun is 81,958,000 miles. The closest Asteroid B is ever to the sun is 12,529,000 miles. The closest Asteroid C is ever to the sun is 92,595,000 miles.

1. To determine if any of the asteroids pass within the range to be classified as an NEO, fill in the missing values from the table.

2. Based on the measurements from Problem 1, do any of the three asteroids qualify as an NEO?

There are two scales that astronomers use to explain the potential danger of NEOs. The Torino Scale is a scale from 0 to 10 that indicates the chance that an object will collide with the Earth. A rating of 0 means there is an extremely small chance of a collision and a 10 indicates that a collision is certain to happen. The Palermo Technical Impact Hazard Scale is used to rate the potential impact hazard of an NEO. If the rating is less than −2, the object poses a very minor threat with no drastic consequences if the object hits the Earth. If the rating is between −2 and 0, then the object should be closely monitored as it could cause serious damage.

⊕ Go to the NASA website http://neo.jpl.nasa.gov/risk/ to answer the following questions.

3. Does any NEO have a Torino Scale rating higher than 0? If so, what is the object's designation (or name) and during which year range could a potential impact occur?

4. Which NEO has the highest Palermo Scale rating? During which year range could a potential impact occur?

Math@Work

Math Education

As a math instructor at a public high school, your day will be spent preparing class lectures, grading assignments and tests, and teaching students with a wide variety of backgrounds. While teaching math, it is your job to explain the concepts and skills of math in a variety of ways to help students learn and understand the material. As a result, a solid understanding of math and strong communication skills are very important. Teaching math is a challenge and being able to understand the reasons that students struggle with math and empathize with these students is a critical aspect of the job.

Suppose that the next topics you plan to teach to your algebra students involve finding the greatest common factor and factoring by grouping. To teach these skills, you will need to plan how much material to cover each day, choose examples to walk through during the lecture, and assign in-class work and homework. You decide to spend the first day on this topic explaining how to find the greatest common factor of a list of integers.

1. It is usually easier to teach a group of students a new topic by initially showing them a single method. If a student has difficulty with that method, then showing the student an alternative method can be helpful. Which method for finding the greatest common factor would you teach to the class during the class lecture?

2. On a separate piece of paper, sketch out a short lecture on finding the greatest common factor of a list of integers. Be sure to include examples that range from easy to difficult.

3. While the class is working on an in-class assignment, you find that a student is having trouble following the method that you taught to the entire class. Describe an alternative method that you could show the student.

4. From your experience with learning how to find the greatest common factor of a list of integers, what do you think are some areas that might confuse students and cause them to struggle while learning this topic? Explain how understanding the areas that might cause confusion can help you become a better teacher.

Math@Work

Physics

As an employee of a company that creates circuit boards, your job may vary from designing new circuit boards, setting up machines to mass produce the circuit boards, to testing the finished circuit boards as part of quality control. Depending on your position, you may work alone or as part of a team. Regardless of who you work with, you will need strong math skills to be able to create new circuit board designs and strong communication skills to describe the specifications for a new circuit board design, describe how to set up the production line, or explain why a part is faulty.

Suppose your job requires you to create new circuit boards for a variety of electronic equipment. The latest circuit board that you are designing is a small part of a complicated device. The circuit board you create has three resistors which run in parallel, as shown in the diagram.

Two of the resistors were properly labeled with their correct resistance, which is measured in ohms. The first resistor has a rating of 2 ohms. The second resistor has a rating of 3 ohms. The third resistor was taken from the supply shelf for resistors of a certain rating, but the resistor was unlabeled. As a result, you are unsure if it has the correct resistance for the current you want to produce. You use an ohmmeter, a device that measures resistance in a circuit, to determine that the total resistance of the circuit you created is $\frac{30}{31}$ ohms.

You know that the equation to determine the total resistance R_t is $\frac{1}{R_t} = \frac{1}{R_1} + \frac{1}{R_2} + \frac{1}{R_3}$, where R_1 is the resistance of the first resistor, R_2 is the resistance of the second resistor, and R_3 is the resistance of the third resistor.

1. Use the formula to determine the resistance of the third resistor given that the total resistance of the circuit is $\frac{30}{31}$ ohms.

2. Was the third resistor on the correct shelf if you took it from the supply shelf that holds resistors with a rating of 7 ohms?

3. What would be the total resistance of the circuit if the third resistor had a rating of 7 ohms?

4. What do you think would happen if the resistance of the unlabeled resistor wasn't determined and the circuit board was sent to the production line to be mass produced?

Math@Work

Forensic Scientist

As a forensic scientist, you will work as part of a team to investigate the evidence from a crime scene. Every case you encounter will be unique and the work may be intense. Communication is especially important because you will need to be clear and honest about your findings and your conclusions. A suspect's freedom may depend on the conclusions your team draws from the evidence.

Suppose the most recent case that you are involved in is a hit-and-run accident. A body was found at the side of the road with skid marks nearby. The police are unsure if the cause of death of the victim was vehicular homicide. Among the case description, the following information is provided to you.

Accident Report	
Date:	June 14
Time:	9:30 pm
Climate:	55 degrees Fahrenheit, partly cloudy, dry
Description of crime scene:	
Victim was found at the side of a road. Body temperature upon arrival is 84.9 °F. Posted speed limit is 30 mph. Road is concrete. Conditions are dry. Skid marks near the body are 88 feet in length.	

Known formulas and data:

A body will cool at a rate of 2.7 °F per hour until the body temperature matches the temperature of the environment. Average human body temperature is 98.6 °F.

Impact Speed and Risk of Death	
Impact Speed	**Risk of Death**
23 mph	10%
32 mph	25%
42 mph	50%
58 mph	90%
Source: 2011 AAA Foundation for Traffic Safety "Impact Speed and Pedestrian's Risk of Severe Injury or Death"	

Braking distance is calculated using the formula $\frac{s}{\sqrt{l}} = k$, where s is the initial speed of the vehicle in mph, l is the length of the skid marks in feet, and k is a constant that depends on driving conditions. Based on the driving conditions on that road for the last 12 hours, $k = \sqrt{20}$.

1. Based on the length of the skid marks, how fast was the car traveling before it attempted to stop? Round to the nearest whole number.

2. Based on the table, what percent of pedestrians die after being hit by a car moving at that speed?

3. Based on the cooling of the body, if the victim died instantly, how long ago did the accident occur? Round to the nearest hour.

4. Can you think of any other factors that should be taken into consideration before determining whether the impact of the car was the cause of death?

Math@Work

Other Careers in Mathematics

Earning a degree in mathematics or minoring in mathematics can open many career pathways. While a degree in mathematics or a field which uses a lot of mathematics may seem like a difficult path, it is something anyone can achieve with practice, patience, and persistence. Three growing fields of study which rely on mathematics are actuarial science, computer science, and operations research. While each of these fields involves mathematics, they require special training or additional education outside of a math degree. A brief description of each career is provided below along with a source to find more information about these careers.

Growing Fields of Study

Actuarial Science: The field of actuarial science uses methods of mathematics and statistics to evaluate risk in industries such as finance and insurance. Visit www.beanactuary.org for more information

Computer Science: From creating web pages and computer programs to designing artificial intelligence, computer science uses a variety of mathematics. Visit computingcareers.acm.org for more information.

Operations Research: The discipline of operations research uses techniques from mathematical modeling, statistical analysis, and mathematical optimization to make better decisions, such as maximizing revenue or minimizing costs for a business. Visit www.informs.org for more information.

There are numerous careers that have not been discussed in this workbook. Exploring career options before choosing a major is a very important step in your academic career. Learning about the career you are interested in before completing your degree can help you choose courses that will align with your career goals. You should also explore the availability of jobs in your chosen career and whether you will have to relocate to another area to be hired. The following web sites will help you find information related to different careers that use mathematics. Another great resource is the mathematics department at your college.

The **Mathematical Association of America** has a website with information about several careers in mathematics. Visit www.maa.org/careers to learn more.

The **Society for Industrial and Applied Mathematics** also has a webpage dedicated to careers in mathematics. Visit www.siam.org/careers to learn more.

The **Occupational Outlook Handbook** is a good source for information on educational requirements, salary ranges, and employability of many careers, not just those that involve mathematics. Visit http://www.bls.gov/ooh/ to learn more.

Answer Key

Chapter 1.R: Number Systems and Fundamental Concepts of Algebra

1.R.1 Exercises

Concept Check

1. False; Equals 81
3. False; 7^0 is 1.
5. True
7. False; 7605 is divisible by 5.
9. False; A prime number has exactly 2 factors.
11. False; 231 is a composite number.
13. False; The LCM of 15 and 25 is 75.
15. True

Practice

17. **a.** 4 **b.** 0 **c.** 1
19. 3, 5
21. None
23. Prime
25. 5^3
27. 24
29. 15

Applications

31. **a.** No. The formula incorrectly shows that each shirt is $10, and the total cost would be $25 − $10 · 11 = $25 − $110 = −$85.

    b. $165; ($25 − $10)·11
33. 4 team members would work 110 hours each; 8

team members would work 55 hours each.

35. 1, 2, 3, 4, 6, 8, 9, 12, 18, 24, 36, 72

Writing & Thinking

37. If addition is within parentheses (or other grouping symbols), addition would be performed first.
39. No, some odd numbers are the product of two or more odd prime factors, for example, 3 · 3 = 9, 3 · 5 = 15, 3 · 7 = 21, etc.
41. Since the LCM is constructed using the prime factors of each number in the set, by definition, each number will divide the LCM.

1.R.2 Exercises

Concept Check

1. True

Practice

3. $\frac{1}{4}$
5. $\frac{1}{3}$

Applications

7. 360 students

1.R.3 Exercises

Concept Check

1. True
3. False; A decimal number that is between 0.01 and 0.10 is between 1% and 10%.
5. False; Fractions that have denominators other than 100 can be changed to a percent.

Practice

7. 20%
9. 2%
11. 0.07
13. 75%
15. $1\frac{1}{5}$

Applications

17. 4%
19. 85%

Writing & Thinking

21. Percent means per centum or per 100. For example, fifty-eight percent means 58 out of 100. Percent can be written as a fraction with 100 in the denominator as in 58/100. The decimal equivalent, 0.58 is read as "fifty-eight hundredths," indicating percent can be written us-

ing the hundredths place, another connection.

23. 100% = 1 so anytime there is a mixed number, which has a value greater than 1, the percentage will be greater than 100%. Proper fractions (numerator is smaller than denominator) have a value less than 1 and therefore the percentage will be less than 100%.

1.R.4 Exercises

Concept Check

1. True
3. True

Practice

5. $9\sqrt{2}$
7. $2x^5 y\sqrt{6x}$
9. $-2x^2 \sqrt[3]{x^2}$

Application

11. $\sqrt{3} \approx 1.73$ amperes

Writing & Thinking

13. A cube root has no restrictions as the cube root of a negative number is negative.

Chapter 2.R: Equations and Inequalities of One Variable

2.R.1 Exercises

Concept Check

1. True

3. False; The statement
$$\frac{1}{3} \cdot \frac{2}{5} = \frac{2}{5} \cdot \frac{1}{3}$$
is an example of the commutative property of multiplication.

5. False; The reciprocal of 1 is 1.
7. False; The reciprocal of 12 is $\frac{1}{12}$.

Practice

9. $\frac{1}{4}$
11. $\frac{1}{3}$

13. $\dfrac{8}{9}$

15. Undefined

Applications

17. $\dfrac{3}{8}$

19. 200 years

Writing & Thinking

21. No. If a fraction is less than 1 then its product with another number will be less than that other number. So, if the other number is less than 1, the product will be less than 1. Answers will vary.

23. $0 = \dfrac{0}{1}$ and the reciprocal would be $\dfrac{1}{0}$ but division by 0 is undefined. So 0 has no reciprocal.

2.R.2 Exercises

Concept Check

1. True

3. False; LCD stands for least common denominator.

5. True

Practice

7. $\dfrac{17}{21}$

9. $\dfrac{23}{42}$

Applications

11. 1 ounce

13. The LCM finds the least common multiple of a set of numbers. The LCD does the same thing for the set of numbers determined by the denominators.

2.R.3 Exercises

Concept Check

1. True

3. False; Odd integers are integers that are not even.

Practice

5. $x - 5 = 13 - x$; 9

7. $n + (n+1) + (n+2) = 93$; 30, 31, 32

Applications

9. **a.** The unknown value is the length of the call in minutes.

b. $m = 20$

c. The collect call lasted 20 minutes.

Writing & Thinking

11. **a.** $n, n+2, n+4, n+6$

b. $n, n+2, n+4, n+6$

c. Yes. Answers will vary.

2.R.4 Exercises

Concept Check

1. False; A proportion is a statement that two ratios are equal.

3. True

5. True

Practice

7. False

9. $B = 7.8$

Applications

11. 180 minutes or 3 hours

Chapter 4.R: Relations, Functions, and Their Graphs

4.R.1 Exercises

Concept Check

1. True

3. True

Practice

5. **a.** 36

b. 16

7. -10

9. 129

Applications

11. **a.** $-\$42 - \$35 - (3 \cdot \$5)$.

b. $-\$92$.

Writing & Thinking

13. Smaller; When any positive number is multiplied by a fraction (or decimal) between 0 and 1, the result will be smaller. This is what is happening when a number between 0 and 1 is squared. Answers will vary.

4.R.2 Exercises

Concept Check

1. True

3. False; In the term "12a," 12 is the coefficient.

Practice

5. -5, 3, and 8 are like terms; $7x$ and $9x$ are like terms.

4.R.3 Exercises

Concept Check

1. True

Practice

3. $15x$

5. $10x^2 + 28x - 76$

7. $8a$

4.R.4 Exercises

Concept Check

1. True

3. False; Subtraction is indicated by the phrase "five less than a number."

Practice

5. $x + 6$

7. $\dfrac{x}{2} - 18$

9. **a.** $4n - 6$

b. $6 - 4n$

11. The product of a number and negative nine

Writing & Thinking

13. The Commutative Property of Addition and Multiplication permits the order of items being added or multiplied to change and still have the same result. This property does not hold true for subtraction or division. Therefore, order is important for subtraction and division problems or the answer will change or be incorrect.

Chapter 5.R: Polynomial Functions

5.R.1 Exercises

Concept Check

1. False; Variables need to be considered as well.

Practice

3. 2

5. $13ab$

Application

7. She can make 30 identical treat bags, each containing 5 pieces of candy A, 6 pieces of candy B, and 11 pieces of candy C.

5.R.2 Exercises

Concept Check

1. False; The middle term should be the sum of the inner and outer products.

3. True

Practice

5. $-(x-2)(x-1)$

7. $(3x+4)(3x-5)$

9. $(5a-2)(a-1)$

Writing & Thinking

11. This is not the completely factored form because $(5x-15)$ can be factored into $5(x-3)$.

5.R.3 Exercises

Concept Check

1. False; The first step is to check for a common monomial factor.

3. False; It might be factorable by the grouping method.

Practice

5. $(x+5)^2$

7. $(5x+6)(4x-9)$

9. $(x+5)(x^2-5x+25)$

Chapter 6.R: Rational Functions and Conic Sections

6.R.1 Exercises

Concept Check

1. True

3. False; Rational numbers have zero denominators.

Practice

5. $\dfrac{3x}{4y}$; $x \neq 0, y \neq 0$

7. $\dfrac{x-3}{y-2}$; $y \neq -2, 2$

Application

9. **a.** $p(x) = \dfrac{15x+200}{x}$

   b. \$35

   c. $x \neq 0$

   d. The variable cannot be negative because you cannot have a negative

quantity of people. There would also be a maximum number depending on the size of the room.

Writing & Thinking

11. **a.** A rational expression is an algebraic expression that can be written in the form $\dfrac{P}{Q}$ where P and Q are polynomials and $Q \neq 0$.

   b. $\dfrac{x-1}{(x+2)(x-3)}$ Answers will vary.

   c. $\dfrac{1}{x+5}$ Answers will vary.

6.R.2 Exercises

Concept Check

1. False; The product will be a binomial.

3. True

Practice

5. $x^2 - 14x + 49$; Perfect square trinomial

7. $9x^2 - 12x + 4$; Perfect square trinomial

Application

9. **a.** $A(x) = 400 - 4x^2$

   b. $P(x) = 4(20-2x)$

Writing & Thinking

11. $(x+5)^2 = x^2 + 2(5x) + 5^2$. Answers will vary.

6.R.3 Exercises

Concept Check

1. True

3. False; The sum of two squares is not factorable.

Practice

5. $(y-8)^2$

Application

7. **a.** $x^2 - 16$

   b. $x-4$ / $x+4$

Chapter 7.R: Exponential and Logarithmic Functions

7.R.1 Exercises

Concept Check

1. False; If there is no exponent written, the exponent is assumed to be 1.
3. True

Practice

5. y^{11}
7. $\dfrac{1}{x}$

Application

9. 2^8 GBs

7.R.2 Exercises

Concept Check

1. True
3. False; The rules for exponents can be applied in any order, resulting in the same answer.

Practice

5. 64
7. $-\dfrac{2y^6}{27x^{15}}$

9. $\dfrac{y^8}{16x^8}$

7.R.3 Exercises

Concept Check

1. True
3. False; If x is a real number, then $\sqrt{x^2} = |x|$

Practice

5. $\sqrt[3]{8}$

7. $\dfrac{1}{10}$
9. $a^{\frac{1}{4}}$

Application

11. 576 ft²

Writing & Thinking

13. No

$$\sqrt[5]{a} \cdot \sqrt{a} = a^{\frac{7}{10}}, \sqrt[5]{a^2} = a^{\frac{2}{5}},$$
$$a^{\frac{7}{10}} \neq a^{\frac{2}{5}}$$

Chapter 8.R: Systems of Equations

8.R.1 Exercises

Concept Check

1. False; The solution must be checked in all equations.
3. True

Practice

5. a, c
7. (4, 0)

Application

9.

5 gallons of 12%; 10 gallons of 3%

Writing & Thinking

11. The solution to a consistent system of linear equations is a single point, which is easily written as an ordered pair.

8.R.2 Exercises

Concept Check

1. False; When boundary lines are parallel, the solution is either the strip between the boundary lines, a half-plane, or there is no solution.
3. True

Practice

5.

7.

Application

9. a. $\begin{cases} 150x + 75y \geq 14{,}000 \\ x + y \leq 150 \end{cases}$

b.

c. Answers will vary.
d. No. $150 \cdot \$75 = \$11{,}250$

Notes

Notes

Notes

Notes

Notes

Notes

Notes

Notes

Notes

Notes